SCIENTIFIC AND EDUCATIONAL JOURNAL OF THERAPEUTIC RIDING

2016/17

ISSN 1174-6238

ISBN 978-0-9973368-1-8

A PUBLICATION OF

THE FEDERATION OF HORSES IN EDUCATION AND THERAPY INTERNATIONAL A.I.S.B.L.

THE FEDERATION OF HORSES IN EDUCATION AND THERAPY INTERNATIONAL A.I.S.B.L

SECRETARIAT
HETI
P.O. Box 134, Damariscotta, ME 04543
Email: office@HETIfederation.org
www.HETIfederation.org
www.facebook.com/HETI.international

ISSN 1174-6238
ISBN 978-0-9973368-1-8

Table of Contents

Dear Readers

I'm happy to share this journal with you. In the late 1980s I heard about Therapeutic Riding from a family friend in Germany. She knew I was a recently graduated physical therapist and a rider, and suggested that I visit Warendorf. During my stay I visited the office of the German Kuratorium (DKThR). I purchased a video and received information about centers where hippotherapy and therapeutic riding was taking place and I ended up visiting many centers around Germany. In the following year, the HETI (formerly FRDI) Congress in Toronto took place, which I attended as a Finnish representative. I spent my month in Canada visiting CanTRA centers in Guelph, Toronto and Ottawa. Since, I have attended and visited as many congresses and centers as possible to get to know the people who work in this field and to learn how they do the work.

Visiting different countries as a guest and as a lecturer I have heard wonderful stories from HETI board members teaching in various centers and spreading the knowledge from their own curricula. With shared information, countries have adopted those parts of the curricula that suit them and the specific regulations of their country.

Again and again I am asked why equine activities and therapies are not more supported. We still lack the research to show cost efficiency and specific patient benefits. Therapies provided as a part of medical treatment require systematic documentation of efficacy and outcomes. If we were able to coordinate our efforts and pool our data, we could start collecting sufficient numbers of data to provide the necessary evidence.

In 2003 researchers and those keen to do research were able to share interests on the HETI website. Unfortunately, that was not yet the time to collaborate internationally. If we could gather the same type of data collected from different countries into a large, international database we could gain more insight into our services and most importantly, gain more support for what we do. The people making decisions in health care and other areas believe in numbers, and through statistical analysis we can speak the language of these decision makers.

I truly believe that now is the time to put our heads together and create evidence-based documentation of our work, and not consider it a novelty anymore. We have accessible and rapid means of communication, training and education, a peer reviewed journal and interest from the general public. At the 16th International Congress of HETI in Dublin next summer, an open forum will be organized with the goal of facilitating discussion and perhaps creating an international consensus around some of the issues discussed in the Editorial.

I want to thank our Editor, Nina Ekholm-Fry, and the peer-reviewers for their extensive work on behalf of the Journal. I also want to encourage you to share your work through manuscript submissions and continued international dialogue.

Sanna Mattila-Rautiainen
President, HETI (The Federation of Horses in Education and Therapy International A.I.S.B.L.)

Editorial

For the past 70 years, the practice of including horses in therapy and education services has evolved, as any emerging field, through organization, education, and regulation. In the late 60s and early 70s, countries like the United Kingdom, Germany, and the United States saw organizations forming with the mission to support therapeutic interactions with horses. The formation of national organizations was, of course, evidence that services were being offered in different parts of Europe and the United States prior (see Ekholm Fry, 2013, for a brief overview).

With organization came better opportunity to educate practitioners and the public. The first international congress was held in Paris in 1974. Six years later, in 1980, the Federation of Riding for the Disabled International, now called Federation of Horses in Education and Therapy International (HETI), was founded to facilitate international collaboration in the field. Countries around the world are in different stages of organization and education, with some relying on the international federation directly, and others having already created strong national organizations that offer support and exchange of information with other countries.

When it comes to regulation, each country must work within its existing professional structures to accomplish this goal. Regulation helps protect both those receiving services and those providing services. It also helps clarify to other professionals and the public what to expect from services, which can help the different kinds of therapeutic and learning services with horses become more accepted worldwide. The United States, part of early development in the field through frequent interactions with European practitioners, now has independent certification boards for professionals such as physical therapists, occupational therapists, speech-language pathologists, psychotherapists, and educators, all of whom include horses in their services; and independent board evaluation for therapeutic riding instructors is in development. Separating education and training from evaluation and certification (or licensing in some cases) is common in many professional fields and helps uphold national standards, especially where there are many opportunities for training. Regulation, in other words, depends on organization (standards and competencies), and education (ways to reach these standards through appropriate training and supervision).

A way for a field to educate and organize is through professional meetings. The 16th international congress, organized by HETI, is held June 25-29, 2018, in Dublin, Ireland. The congress promotes sharing of information, exchange of ideas, and a feeling of connection and support, no matter where you live. When I attended my first HETI congress over a decade ago, it was the first time I had been to a meeting of that scale in our field. Being surrounded by colleagues from all over the world with the same professional interests was and still is an amazing experience.

Regardless of country or stage of organization, there are a few issues we appear to have in common: terminology, professional scope, and equine welfare. Large variations in use of terms to describe what service we provide are not only an issue in comparing research studies and in international communication, but also for practitioners, clients, and third parties such as health insurance providers. This relates closely to the second issue: professional scope. Clarity is needed around who is qualified to deliver a service with horses in therapy or education, that is, what kind of education, training and credentialing is necessary to provide safe, ethical, and effective services, along with the practitioner's ability to provide detailed and accurate descriptions of the services they deliver in order to avoid misrepresentation. The worst-case scenario in any professional field is when the public is lead to trust individuals who are not qualified to deliver the services they advertise and both clients and horses are endangered, physically and mentally.

This leads us to the third area of challenge: equine welfare and attitudes towards horses in our field, and in general. The inclusion of horses in services for humans has the potential to just be another way we exploit other animals, whether we wish to think that we are helping both humans and horses in our professional pursuits or not. In order to avoid this, and to make our services truly ethical and sustainable, our knowledge about horses must be based on the best available scientific information about their behavior and needs. Especially relevant is how we house them, and how we train and handle them. Having accurate knowledge about how horses think and their emotional and physical needs might mean that we have to change or modify what we do in order to create the best environment for all, not just for humans receiving the service. The interconnectedness of human health, animal health, and environmental health (One Health), can help us create a framework for this way of thinking.

A Letter to the Editor in this issue addresses problems in a process that helps a field educate itself: publication in a peer-reviewed academic journal. The reader of a peer-reviewed journal, such as the one in which the article addressed in the Letter was published, will expect that claims in such articles be based on scientifically sound information, unless otherwise noted. The article by Usadi and Levine (2017) unfortunately contains a number of unsupported statements that are presented as facts. There could be several reasons why these issues were not addressed in the peer-review process (a process which constitutes best practice when it comes to publication of scientific literature). The Scientific and Educational Journal of Therapeutic Riding engages a team of peer-reviewers with expert knowledge in a variety of equine-assisted services, which creates a thorough review process. Reasons for the issues addressed in the Letter and the journal in which the article in question was published could have been the lack of knowledgeable, subject-specific reviewers and editors, and an overall inappropriate forum for the nature of the text the authors submitted.

In this issue of the Journal:

- Goldberg and colleagues evaluated the impact of an equine-assisted learning program on mindfulness in medical students
- Bassette and colleagues presented a study on increasing conversational skills in adolescents during therapeutic riding through three cases
- Loukas and colleagues studied experiences by riders, parents and volunteers at a therapeutic riding program from an occupational therapy theoretical framework
- Mirzabeigi and colleagues studied the effect of hippotherapy on static and dynamic balance, and reaction time, for autism spectrum disorder.

In addition, there is a Special Commentary from the International Society of Equitation Science (ISES) 2016 conference in Saumur, France by Heta Rautiainen and myself. Accurate and scientifically based knowledge about horses is necessary for equine-assisted practitioners which makes information from groups like ISES of great benefit to our field.

As always, please direct your emails with comments and questions to editorial@hetifederation.org

In the spirit of support and encouragement (and hope to see you in Dublin!),

Editor
Nina Ekholm Fry (USA/Finland)
Institute for Human-Animal Connection
University of Denver

References

Ekholm Fry, N. (2013). Equine-assisted therapy: An overview. In M. Grassberger et al (Eds.). *Biotherapy: History, principles and practice* (255-282). London: Springer

Usadi, E. J., & Levine, R. A. Why we don't ride: Equine assisted psychotherapy, military veterans and moral injury. *Journal of Trauma Treatment*, 6(2) doi: 10.4172/2167-1222.1000374

Call for Manuscripts

The Federation of Horses in Education and Therapy International (HETI) publishes the Scientific and Educational Journal of Therapeutic Riding (SEJThR) annually. We are currently seeking submissions with the purpose of advancing knowledge in the fields of therapeutic riding and equine-assisted activities and therapies.

The SEJThR has several submission categories:

- **Original Research**
  An original research study conducted by the author
- **Review**
  A comprehensive or systematic review of research on a specific topic
- **Case Study**
  A presentation of a single comprehensive case study or multiple case studies
- **Theory and Practice**
  A discussion of a specific topic in the field
- **Letter to the Editor**
  Comments on articles published in the SEJThR or in other journals on topics related to the equine-assisted field

Manuscripts are reviewed through a double-blind peer-review system and submissions are welcome throughout the year.

For more information about manuscript requirements and the submission process, please visit www.hetifederation.org/journal.html.

Please contact journal editor Nina Ekholm Fry with questions at editorial@hetifederation.org.

Journal Editor:
Nina Ekholm Fry
editorial@hetifederation.org
www.hetifederation.org

Call for Reviewers

The Federation of Horses in Education and Therapy International (HETI) publishes the Scientific and Educational Journal of Therapeutic Riding (SEJThR) annually. The focus of this journal is on advancing knowledge in the fields of therapeutic riding and equine-assisted activities and therapies.

Reviewers are needed to ensure the quality and status of the SEJThR as a peer-reviewed journal. Peer-reviewers conduct blind reviews of manuscript submissions within their area of knowledge as a step in the review process.

Expectations for reviewers are:

- Ability to provide constructive feedback for a scholarly manuscript in terms of its quality and contribution to the field
- Timely and respectful communication with journal editors

Requirements for reviewers are:

- Professional experience in an area of therapeutic riding/equine-assisted activities and therapies
- A master's or doctoral degree
- Experience in reviewing scholarly texts
- Ability to complete a review within four weeks of receiving a manuscript

If you are interested in contributing your time as a peer-reviewer for the SEJThR, please submit a letter of interest and your curriculum vitae (CV) per email to editorial@hetifederation.org. Please type "Reviewer Application" in the subject line.

Journal Editor:
Nina Ekholm Fry
editorial@hetifederation.org
www.hetifederation.org

LETTER TO THE EDITOR

A Response to Usadi and Levine (2017), "Why We Don't Ride: Equine Assisted Psychotherapy, Military Veterans and Moral Injury

Jason R. Ledlie

Correspondence to:
Jason R. Ledlie
PO Box 444
Parshall, CO 80468
Email: JasonLedlie@gmail.com
Phone: (512) 621-9684

KEYWORDS

Equine-related therapies, Research, Scientific Evidence, Models.

A Response to Usadi and Levine (2017), "Why We Don't Ride: Equine Assisted Psychotherapy, Military Veterans and Moral Injury

I am a mental health professional who incorporates equine-assisted work into my clinical practice in the United States. I am writing to implore you to entertain a critique of the recently published article by Usadi and Levine (2017).

Foremost, I am compelled to express that it is not my aim to be provocative, and yet I feel like the integrity of the equine-assisted field deserves an expanded dialogue. I'd also like to clearly state that my intention is not to detract from the work being conducted at Trauma and Resiliency Resources, Inc., nor to draw into question the effectiveness of that work.

Being published is always a noteworthy achievement, and I commend Usadi and Levine for their effort. The equine-assisted body of literature does not enjoy the breadth many would prefer, and so these pursuits merit recognition. However, it occurs to me that the article in question was published in a peer-reviewed scientific journal without purporting to boast any scientific findings to support the claims found therein. At the risk of offending, my fear is that this only fuels our detractors, who will use the publication of unsupported statements in their quest to undermine the field.

In their article, Usadi and Levine (2017) submit argument for why working from the ground in equine-assisted psychotherapy (EAP) is thought to be superior to, or is preferred at the exclusion of, mounted work (i.e., someone being on a horse). I take no exception to practitioners who elect to work exclusively from the ground. However, the rationale for doing so detailed in this article appears near-sighted and lacks scientific substantiation.

As an example, in their abstract, Usadi and Levine (2017) state that the model of EAP utilized in their program is the most ideal for warriors suffering moral injury, citing the horse's ability to exhibit "nonjudgmental intuitive mirroring." This statement is further qualified by noting that "a horse's ability to intuitively target post-war attachment disruptions caused of morally injurious combat experiences is best employed when the veterans keep their feet on the ground" (Usadi & Levine, 2017). I do not bring this up to argue their point; perhaps that is very well the case. However, I find it difficult to reconcile the notion that a creature able to "intuitively target post-war attachment disruptions" is simultaneously unable to, intuitively or otherwise, have an accurate appraisal of the threat level of being saddled for mounted work (which is used as an argument against using mounted work for war veterans). This seems especially true when we consider learning theory and the idea that saddling should be a well habituated experience for horses (see Doner & Ekholm Fry, 2015).

While I applaud Usadi and Levine (2017) for recognizing the importance of addressing moral injury, I do not subscribe to the notion that moral injury must be addressed at the exclusion of trauma and traumatic stress responses. Usadi and Levine (2017) clearly state that "working with horses on the ground and while riding can serve to regulate the nervous system of the post-war combat veteran and that both modes of horse work can begin the healing of post-war attachment disruption." With symptoms related to trauma exposure having a tendency to be quite pervasive, one could argue that you cannot work meaningfully in a therapeutic setting without addressing those trauma responses, even if secondarily. If we agree that the benefits of mounted movement have a desirable effect on the nervous system under such circumstances, then omitting the possibility for mounted work only seems limiting. To their point about "correcting" or "instructing" having the potential to be counterproductive in a psychotherapy setting, I agree. That said, their assertion that mounted work must include horsemanship instruction is roundly false. While I won't expound on examples in the interest of brevity, there are simple and functional methods for realizing the benefits of mounted work and equine movement, especially in time-limited settings, that require little to no skill on part of the client, and this can still be achieved while observing ethical considerations for both client and equine.

Lastly, Usadi and Levine (2017) plead the case for the Eagala model and its operational requirement consisting of a licensed mental health professional (MH) and equine-specialist (ES) team. Their points regarding the importance of the involvement of a licensed mental health professional are sound since the purpose, in this case, is treatment (psychotherapy). I would submit, however, that the structural integrity of the equine-assisted field would be more robust were dual competency the expectation for equine-assisted clinicians; that is to say, they should be equally knowledgeable therapists and equestrians, able to practice unaccompanied (noting that some settings may require the assistance of others).

With much infighting and posturing for superiority amongst the varying equine-assisted models, some objective readers may get the impression that this article is little more than a veiled attempt to justify the continued commerce of the model in question, and that it is not meant to scientifically substantiate any benefit of equine-assisted intervention. The latter is what the field needs.

References

Doner, E., & Ekholm Fry, N. (2015). Role of learning theory in training and handling the therapy horse. *Scientific and Educational Journal of Therapeutic Riding*, 20, 61-76

Usadi, E. J., & Levine, R. A. (2017). Why we don't ride: Equine assisted psychotherapy, military veterans and moral injury. *Journal of Trauma Treatment*, 6(2) doi: 10.4172/2167-1222.1000374

Special Commentary

International Society for Equitation Science Conference, June 2016, Saumur, France
Heta Rautiainen & Nina Ekholm Fry

The International Society for Equitation Science (ISES) is a group dedicated to the scientific understanding of horses and their welfare, so that the quality of their association with humans can be improved. By supporting research into the training and welfare of horses, and creating an international forum for scientists to communicate across disciplines, ISES encourages the assimilation of scientific knowledge for use in relation to practical problems concerning the way horses are trained, managed, housed and cared for. The society provides free educational content, conferences, and position statements relevant to horse interactions in any situation, which makes awareness of this group relevant to professionals who include horses in therapy and educational services.

A highly recommended resource from ISES is their "First Principles of Horse Training", available for free on their website both as a printable poster and as a text with more in-depth explanations. These ten training principles describe how to maintain optimal welfare and training efficiency regardless of your horse's age, breed, training level, and discipline. The principles apply to training and handling of therapy horses, and you can assess whether your existing training system demonstrates each principle.

There is no reason why professionals who provide equine-assisted services should not be among the most knowledgeable on horse behavior, ethical training, and welfare. It is important to base our understanding of horses on scientific inquiry, rather than simply on tradition or personal experience. Groups like ISES can help us achieve this. In addition to the material available on their website, ISES organizes an annual conference in difference parts of the world, which is open to anyone interested in attending. What follows is a conference report from the 12th ISES conference by Heta Rautiainen (Finland).

In June 2016, the 12th annual International Society for Equitation Science (ISES) conference took place at the French National Riding School in Saumur, France, and was hosted by the French Institute of Horse and Riding (Ifce). The Cadre Noir of Saumur, renowned for its centuries of horse training and cavalry tradition, has its home at the National Riding School, and participants viewed demonstrations as part of the conference.

The 12th conference theme, “Understanding horses to improve training and performance”, is relevant to professionals in all sectors that involve interaction with horses. Demonstrations highlighted the importance of welfare for both horse and rider/handler, as good performance, regardless of setting, is dependent on welfare. One of the key messages from the conference was that in order to improve welfare, scientific study should be applied to all areas of equine interaction. The conference provided new insights in areas of performance (physiology, biomechanics, equipment, impact of surfaces), cognition and learning, behavioral needs, breeding, handling of horses, and human-horse relationships. Presentations generated interesting and open-minded discussions.

The conference started with an introductory overview on what is known about the behavior, needs, and cognition of horses (McDonnell), followed by a thought-provoking presentation in which memory processing between humans and other animals was compared (Calandreau). Understanding the functional anatomy of the brain is very important for making sense of the mechanisms that steer our own and our horses' behaviors. Our communication with horses is based on applied ethology, meaning that we apply information about horses' natural behavior and learning to horse care and training. Understanding equine behavior and equine learning theory is key to improving the human-horse relationship, and also has major benefits for the overall wellbeing of horses. In a study examining four equine-assisted therapy and learning models in the U.S., the authors concluded that more accurate understanding of equine behavior is needed to fully utilize behavioral feedback and to improve welfare for the horses working in this area (Kieson & Abrahamson). Good quality therapy and learning services can only be produced with well-kept and relaxed animals.

Many studies presented at the conference focused on how stress or discomfort from housing conditions and equipment (tack) affect horses' stress levels and, thus, overall behavior. The quality of a horse's performance, regardless of discipline or work role, is a result of both their needs being met in their living situation and appropriate training and equipment use. To improve equine wellbeing we must know what is normal behavior and why problematic behaviors occur. For example, feeding and overall management strategies can have a huge impact on a horse's stress levels, which also closely affect the work we do with the horse. Horses need approximately 16 hours of grazing per day (near-constant intake of fiber), space to move without restriction, and to socialize with other horses. If these basic needs are hindered, horses develop stereotypies, which can be compared to addiction behaviors in human terms (van Dierendonck; Losonci, Berry & Paddison). Stereotypic behaviors arise as a way for horses to cope with their environment. Studies are exploring whether they have an impact on learning (Briefer, Beuret, Briefer, Zuberbühler, Bshary & Bachmann).

Stress, pain, and discomfort can be recognized in horses, even if they, as prey animals, tend to hide such expression. Van Dierendonck and van Loon introduced a study in which two pain scales were validated. Recognizing pain expression can be difficult, so these validated pain scales are much needed in the process of evaluating behavioral problems and welfare issues. In addition to traditional observational methods, a new technological device has been developed to objectively and reliably measure things such as nose band tightness (Casey, Conway, Doherty & Conway), which is a remarkable problem in equine sport (Doherty, Casey, McGreevy & Arkins), as it produces negative effects on equine mental states (Piette, Exadaktylos & Berckmans). In addition, an advanced electroencephalography (EEG) helmet for measuring brain waves in horses was presented. Its ease-of-use will be helpful in studying cognition, monitoring attention during veterinary procedures, and detecting diseases like epilepsy.

The impact of tack and equipment was discussed extensively at the meeting. Pressure and its use in handling and training is an important area to study, as incorrect or inappropriate use of pressure has welfare consequences for all horses. One study (Carey) compared bitted and bitless bridles used in a therapeutic riding setting. Horses showed less negative behavior like head tossing, resistance, biting at the leader, and having a restless mouth when wearing a bitless rather than a bitted bridle in the test group. In a demonstration session, Jill Carey from Festina Lente Equestrian Centre (Ireland) shared her experiences of applying ethology (the study of behavior) to her center environment and educating staff on how a comfortable, low-stress life can be ensured for riding school and therapy horses.

In addition to training and housing, temperament is an important factor to consider in our interactions with horses. Temperament, which is considered fairly fixed, innate, and observable soon after birth, affects how horses react to stimuli. Horses are sensitive animals and a constant state of fear impairs their quality of life and ability to work. Experience of fear-inducing situations, such as novel object exposure, raises heart rate and slows the performance of tasks (Winther-Christensen). Glucocorticoid secretion related to this kind of stress can cause health issues in the long run. Fear and discomfort should be minimized during training in order to avoid stress and unwanted emotional associations to the interactions and tasks involved. Thus, for safety and welfare reasons, horses should be prepared to experience as many things as possible with humans, and this, of course, includes therapeutic and educational services. Helping horses become habituated to different kinds of new objects and situations with a familiar handler (Cliffe &

Scofield), and giving them a chance to associate a relaxed mind to as many things as possible, are good strategies.

The usability of personality tests (Lansade & Vidament) for horses as part of breeding and leisure horse grading systems, in order to objectively categorize "shy" and "confident" horses as well as matching horses with riders, was also discussed at the conference. These kinds of tests need to be carried out in a way that is safe for everyone involved. One particular issue revolves around provoking too much fear and flight during testing, and, thus, causing psychological trauma for horses since, with their large amygdala, they have excellent memory for fear. Applying the principles of learning theory during testing and training can help horses maintain their natural curiosity while remaining calm and safe. Temperament testing prior to specific training and preparation can help establish overall suitability of horses for particular kinds of therapy or learning services.

Understanding interactions between humans and horses is very important whether you are a leisure rider or a professional. Because horses associate feelings with what they experience, every person the horse meets teaches the horse something about humans. There is still a significant number of people who regularly interact with horses and who lack information about approaches to equine management that are based on the horse's behavior and needs (Telatin, Baragli, Greene & Gardner; Bornmann). Fortunately, successful approaches to increasing rider knowledge through equitation pedagogy have been implemented (Abbey & Randle). Experiential learning techniques and ways for learners to find trust in the teacher's competency were impactful aspects for changing behavior and thinking (Dumbell & Lewis).

In addition to the handful of studies highlighted here, the conference contained a variety of very interesting presentations, such as the impact of arena surfaces on horses, pressure and consequences from using specific kinds of equipment, and the biomechanics of both humans and horses. Although this was a scientific conference with serious topics, the atmosphere was uplifting and people very friendly. As professionals working with horses in therapy and education, we should apply scientific knowledge of horses, riding, and human-horse interaction to our practice in order to provide more beneficial and high quality services for our clients, and a better life for our horses.

Information about ISES, upcoming conferences, and proceedings from previous conferences can be found at www.equitationscience.com

Article One

Equine-Assisted Learning (EAL) Improves Mindfulness in Medical Students

Madeline Goldberg, MS
Amanda Gonzalez
Bertha Ben Khallouq, MS
Manette Monroe, MD Med

University of Central Florida College of Medicine

Correspondence to: Madeline.Goldberg@Knights.ucf.edu

Abstract

PROBLEM: Cultivating mindfulness and empathy skills in physicians can reduce medical error, improve accuracy of diagnoses, and improve patient compliance (Beach, et al., 2013). There is a debate among medical educators as to whether these skills can be taught.

INTERVENTION: The Medicine and Horsemanship (M&H) program (Kane, 2007) uses equine-assisted learning (EAL) to teach emotional wellbeing and communication skills to medical students. This program is being used at the University of Central Florida College of Medicine (UCF COM). For this study, all participants completed pre- and post-intervention surveys including: Five Facet Mindfulness Questionnaire (FFMQ) and Experiences Questionnaire (EQ), which measured mindfulness, while the Interpersonal Reactivity Index (IRI) measured empathy. In addition, the intervention group completed a qualitative survey and debrief data was collected.

OUTCOME: Medical students who completed the M&H program had significantly higher post-intervention mindfulness scores in the subscales of nonreactivity and decentering, compared to controls. The M&H program is an innovative method for improving mindfulness among medical students.

LESSONS LEARNED: With ten students per cohort, several iterations of the program would be necessary for significant analysis of outcomes. Due to the relatively small number of volunteers, participants could not be randomly assigned. Analysis of baseline differences did not show significant differences between intervention and controls, adding strength to our conclusions.

Keywords

Mindfulness; Medical Education; Equine-Assisted Activities; Communication; Nonreactivity.

INTRODUCTION

Mindfulness can be defined as quality of mind where one is attentive to oneself, the present moment, and one's surroundings in a non-judgmental manner (Bishop et al., 2004). Research has shown that the practice of mindfulness has associated benefits such as increasing self-esteem (Rasmussen & Pidgeon, 2011) and attention (Sigsbee & Bernat, 2014) as well as decreasing social anxiety (Rasmussen & Pidgeon, 2011). In the medical field, research has found that physicians who practice mindfulness techniques are more successful in communicating with patients (Beach et al., 2013) and that it is possible to teach mindfulness skills to physicians with effects that persist over time (Ammentorp, Sabroe, Kofoed, & Mainz, 2007, Beckman et al., 2012).

Interpersonal and professional communication skills are important skills to be mastered by medical students and can be taught through a variety of methods including experiential learning. For example, one experiential learning technique commonly employed by medical schools is to use standardized patients (patient actors) to teach clinical skills in standardized interactions. This allows medical students to develop skills in a professional and clinical environment. This study assesses whether the Medicine and Horsemanship (M&H) program, developed by Beverley Kane (Kane, 2007), might support traditional and clinical approaches to medical education by teaching mindfulness skills and/or enhancing empathy in medical students through the use of equine-assisted learning (EAL).

Previous research has found that horses can help teach a variety of personal and interpersonal skills. For instance equine-assisted work has been used with war veterans to teach coping strategies for Post-Traumatic Stress Disorder (PTSD) (Nevins, Finch, Hickling, & Barnett, 2013). In addition, horses have successfully been part of teaching autistic children how to better connect with their surroundings (MacLean, 2011). EAL uses horses to promote experiential learning and equine-assisted therapy (EAT) has been found to be effective in improving mental health and overall well-being in participants (Masini, 2010).

There is a need for increased mindfulness as well as empathy in medical professionals. Studies have demonstrated that patient satisfaction positively correlates with physician communication skills (Bishop et al., 2004). In addition, studies have found that physician mindfulness training can reduce medical error, improve accuracy of diagnoses as well as improve patient compliance (Bishop, Lau, Shapiro, Carlson, Anderson, Carmody et al., 2004). The M&H Program could be particularly valuable for medical students who are under significant pressure—both externally imposed by medical institutions as well as imposed by themselves (Masini, 2010).

This study is a continuation of previous research conducted at UCF COM that measured medical student changes in mindfulness, social self-efficacy, nonverbal communication, and stress and depression (Priesthoff, 2015, Kostamo, 2015, Patel, 2015, Gonzalez, 2016, Vaizer, 2014). Although previous M&H research at UCF COM failed to capture significant changes in first-year medical student mindfulness after completing the M&H Program (Priesthoff, 2015), it is hypothesized that the survey tool previously used to measure mindfulness, the Mindful Attention Awareness Scale (MASS), was not appropriate for the M&H Program. A systematic review of the MASS found that the MASS captured significant differences in mindfulness between experienced meditators and non-meditators, but failed to capture differences between those new to meditation and non-meditators (controls). MASS may be less specific for finding changes in mindfulness between those who are new to mindfulness techniques and controls. In the previous study, the M&H Program took place during a consecutive 10-day period in the beginning of summer. In this study, the M&H sessions took place during the academic year (while classes were in session) with six M&H sessions occurring over three months. In addition, this study uses new survey instruments to measure mindfulness, including the Five Facet Mindfulness Questionnaire (FFMQ) (Nevins, Finch, Hickling, & Barnett, 2013) and the Experiences Questionnaire (EQ) (MacLean, 2011), and adds a new variable, empathy, that is measured by the Interpersonal Reactivity Index (IRI) (Davis, 1983). We hypothesize that through the use of new survey instruments and expansion of the intervention time frame, we will find positive changes in mindfulness and/or empathy in medical students who complete the M&H Program.

Methods

Participants

Participants were over 18 years old and enrolled at UCF COM as first-year medical students. Participation was voluntary and recruitment took place through email and interest group meetings. Participants self-selected into either the intervention group, which participated in the M&H Program, or the control group, which did not participate in the M&H Program. Although not ideal, this self-selection was necessary to achieve an adequate number of participants. Twenty (intervention, n=9; controls, n=11; 55% female) volunteered to participate in this study. All 20 participants received a $10 incentive for participating in the study. Data was de-identified to maintain subject confidentiality. This study was approved by UCF COM IRB: SBE-14-10236.

Participant Safety

Prior to beginning the first session, the intervention group received a brief safety orientation at the equestrian facility where the study took place. To ensure participant safety and to facilitate M&H activities, a Professional Association of Therapeutic Horsemanship International (PATH Intl.) certified instructor attended each session. All horses used in the M&H Program were therapeutic riding horses.

M&H Program

M&H sessions were distributed over a three-month period during the academic calendar. All M&H sessions were held at a local therapeutic riding center, Heavenly Hoofs, Inc., in St. Cloud, Florida and were 3 to 4 hours in length. Sessions were run by the research team consisting of two first-year medical students and the principal investigator (a UCF COM physician and faculty member with extensive equine experience).

The M&H Program uses EAL activities that provide participants with opportunities to react to novel situations and reflect upon these experiences. Many of the sessions included activities that served as metaphors for physician-patient interactions. For example, one exercise simulated the patient exam by encouraging participants to be mindful of their environment and reflect on how they were affecting the horses during a veterinary exam. The veterinary exam included measuring heart rate, pupil dilation, and capillary refill (using the horses' gums). Another M&H exercise symbolized guiding and supporting a nervous patient through his or her fears relating to a difficult procedure by encouraging a horse to cross an intimidating obstacle.

The exercises used in the M&H Program further promote interpersonal development of mindfulness and empathy skills. One task, for instance, required group placement of a halter on a horse with the added challenge of only one individual speaking while maintaining physical contact (linking arms) with fellow participants. This exercise teaches the importance of effectively communicating with colleagues by establishing roles and duties and encouraging teamwork. Descriptions of all 6 M&H sessions can be found in Appendix D. Table 1 displays M&H Program objectives listed in the M&H manual (Kane, 2007).

Table 1: M&H Program Objectives

1.	Become aware of subtleties of verbal and nonverbal communication
2.	Improve attention, mindfulness, and focusing abilities
3.	Become aware of incongruency of intention vs. behavior
4.	Identify and respect boundaries in ourselves and others
5.	Recognize the nature of projection and transference
6.	Confront insecurities and develop confidence
7.	Adjust to the relativity of time, expand the moment
8.	Cope with stress

Table 1: M&H Program Objectives (Kane, 2007). Note: The present study uses quantitative and qualitative instruments to measure changes in mindfulness and empathy that are concepts related to many of the objectives above.

DATA COLLECTION

This study used a mixed methods design with both quantitative and qualitative data to examine changes in mindfulness and empathy in medical students. Participants in control and intervention groups completed a confidential survey, including, three psychological instruments: Five Facet Mindfulness Questionnaire (FFMQ); Experiences Questionnaire (EQ); and Interpersonal Reactivity Index (IRI). Pre and post-intervention (one week before and after M&H), surveys were electronically administered through the Qualtrics platform (*Qualtrics* [computer program], 2015).

Qualitative Data Collection. *Debrief Sessions and Qualitative Survey*. Qualitative data was collected in the form of debrief sessions and a qualitative survey. Research-team guided debrief sessions, 30-60 minutes in length, took place at the end of all six M&H sessions. Debrief sessions were transcribed and coded by two researchers using a grounded theory approach (Merriam, 2009). Through this approach, researchers identified nodes of interest—"themes"—that help describe, understand, and interpret the intervention group responses to questions posed during the debrief sessions. Qualitative survey results were analyzed by sight for common themes. See Table 2 for the questions included in the qualitative survey.

Table 2: Qualitative Survey

1. Did you notice a change in your mood/outlook on life the week after participating in the Medicine and Horsemanship Program?
2. Did you feel that participating in the Medicine and Horsemanship Program impacted your approach to handling situations where you did not know exactly what to do?
3. Before beginning the Program, what thoughts/feelings did you have about caring for patients with whom you have little ability to communicate due to language or physical barriers?
4. After completing the Program, what thoughts/feelings do you have about caring for patients with whom you have little ability to communicate due to language or physical barriers?
5. Has participating in the Medicine and Horsemanship Program affected the way you interact with and communicate with your colleagues? If yes, how so?
6. Would you recommend this Program to fellow classmates? If so, why?
7. What was your favorite activity during the Medicine and Horsemanship Program?
8. What was your least favorite activity during the Medicine and Horsemanship Program?
9. What suggestions would you make to improve the Program?

Table 2: Qualitative Survey. The qualitative survey was completed by the intervention group within one week of finishing M&H. The qualitative survey was sent with the post-intervention survey which measured quantitative changes in mindfulness and empathy in intervention group and controls.

Quantitative Data Collection. *Five Facet Mindfulness Questionnaire (FFMQ; 39-items).* The Five Facet Mindfulness Questionnaire (FFMQ) uses five subscales to conceptualize mindfulness including: Observing of Inner Experience (FFMQ OBS); Describing Experience (FFMQ D); Acting with Awareness (FFMQ AA); Nonjudging of Experience (FFMQ NJ); and Nonreactivity to Inner Experience (FFMQ NR). See Appendix A for complete FFMQ instrument (Baer, Smith, Hopkins, Krietemeyer & Toney, 2006). Examples from the FFMQ subscales are below:

1. FFMQ OBS: I pay attention to sensations, such as the wind in my hair or sun on my face.
2. FFMQ D: I can easily put my beliefs, opinions, and expectations into words.
3. FFMQ AA: It seems I am "running on automatic" without much awareness of what I'm doing.*
4. FFMQ NJ: I criticize myself for having irrational or inappropriate emotions.
5. FFMQ NR: I watch my feelings without getting lost in them.

Participants rank their responses on a 5-point Likert scale ranging from 1 = "never or very rarely true" to 5 = "always true". Nineteen items were reverse-scored, including the FFMQ AA example above (see asterisk).

Cronbach's alpha values less than 0.70 were excluded from final data analysis (see Table 3 for Cronbach's alpha values).

Table 3: Instrument Reliability

	Pre-Cronbach's α	**Post-Cronbach's α**
FFMQ		
Observing of Inner Experience (OBS)	0.56	0.88
Describing Experience (D)	0.89	0.93
Acting with Awareness (AA)	0.93	0.95
Nonjudging of Experience (NJ)	0.94	0.92
Nonreactivity to Inner Experience (NR)	0.73	0.80
Total	0.88	0.92

Table 3: Instrument Reliability – cont'd.

	Pre-Cronbach's α	Post-Cronbach's α
IRI		
Perspective Taking (PT)	0.61	0.66
Fantasy (F)	0.68	0.44
Empathic Concern (EC)	0.63	0.42
Personal Distress (PD)	0.80	0.79
EQ		
Decentering (D)	0.83	0.88
Rumination (R)	0.65	0.67

Table 3: Instrument Reliability. Note: Five Facet Mindfulness Questionnaire (FFMQ); Inter-personal Reactivity Index (IRI); Experiences Questionnaire (EQ). Subscales excluded from final analysis due to low instrument reliability included: FFMQ OBS; IRI PT; IRI F; IRI EC; and EQ R.

In this study, FFMQ OBS did not meet the 0.70 threshold; Cronbach's alpha ranges for the remaining FFMQ subscales were from 0.73 to 0.95, thus the internal consistency of instruments used in this study is reliable. Our instrument reliability is similar to that reported in the literature for the FFMQ mindfulness instrument with values ranging from 0.74 to 0.94 (Park, Reilly-Spong & Gross, 2013).

The Experiences Questionnaire (EQ; 20-items). The Experiences Questionnaire (EQ) measures mindfulness as it relates to Decentering (EQ D) and Rumination (EQ R). See Appendix B for complete EQ instrument (Fresco et al., 2007). EQ instrument subscale examples are below:

1. EQ D: I am better able to accept myself as I am.
2. EQ R: I think over and over again about what others have said to me.

Respondents rate their responses on a 5-point Likert scale from 1 = "never" to 5 = "all the time". Developers of the original EQ instrument excluded question two from the 20-item EQ instrument due to very low and unreliable factor loadings (0.12 and 0.18) (Fresco et al., 2007). We followed the recommendations of the original EQ authors and excluded question two from final analysis.

EQ authors further excluded two additional questions from EQ D subscale (due to poor factor loading) and one question from the EQ R subscale (due to double-loading) (Fresco et al., 2007), however, after assessing our own data, we chose to include all questions originally intended to measure decentering and rumination since Cronbach's alpha levels were higher with the questions included. Our final EQ instrument included 13 questions loading onto the EQ D subscale and 7 questions loading onto the EQ R subscale, however, instrument reliability values for EQ R remained below threshold (0.70) and thus EQ R was excluded from final data analysis. The reported Cronbach's alpha for the EQ instrument is between 0.83 and 0.90. In this study, Cronbach's alpha for EQ D ranged from 0.83 to 0.88, thus the internal consistency in this study is reliable.

Interpersonal Reactivity Index (IRI; 28 items). The IRI measures empathy according to four subscales: Perspective Taking (IRI PT); Fantasy (IRI F); Empathic Concern (IRI EC); Personal Distress (IRI PD). See Appendix C for complete IRI instrument (Davis, 1983). IRI subscale examples are below:

1. IRI PT: When I'm upset at someone, I usually try to "put myself in his shoes" for a while.
2. IRI F: I really get involved with the feelings of the characters in a novel.
3. IRI EC: I often have tender, concerned feelings for people less fortunate than me.
4. IRI PD: I sometimes feel helpless when I am in the middle of a very emotional situation.

Respondents rate items on a 5-point Likert scale with values ranging from 0 = "describes me very well" to 5 = "does not describe me very well". Nine items were reverse-scored. IRI PT, FS, and EC were excluded from final data analysis due to low instrument reliability. Only IRI PD was included in final analysis as Cronbach's alpha values exceeded the 0.70 threshold with values ranging from 0.80 to 0.79.

Data Analysis

Statistical analyses were performed on pre- and post-intervention surveys using mixed repeated measures analysis of variance (RM-ANOVA) to evaluate first, baseline differences between control and intervention groups and second, any changes in participants' empathy and mindfulness which may have emerged over time. Continuous data are expressed as mean ± standard deviation (see Table 4).

Table 4: Pre- and Post-Intervention Mindfulness and Empathy Subscales

	Control (n=11) Mean (SD)	Intervention (n=9) Mean (SD)
FFMQ: Five Facet Mindfulness Questionnaire		
FFMQ D: Describing Experience		
pre	26.09 (4.25)	23.89 (5.62)
post	25.82 (6.51)	24.56 (3.57)
FFMQ AA: Acting with Awareness		
pre	24.00 (6.93)	25.44 (6.97)
post	25.45 (6.44)	25.56 (6.33)
FFMQ NJ: Nonjudging of Experience		
pre	23.18 (8.27)	25.67 (6.02)
post	24.27 (6.66)	25.11 (6.01)
FFMQ NR: Nonreactivity to Inner Experience		
pre	22.64 (3.56) ±*	21.11 (4.28) ±*
post	22.82 (3.71) ±*	24.11 (3.76) ±*
FFMQ Total: Sum of FFMQ Subscales		
pre	120.82 (16.50)	122.44 (16.74)
post	123.73 (17.27)	130.11 (17.93)
EQ: Experiences Questionnaire		
EQ D: Decentering		
pre	41.36 (6.85) ±□	39.88 (4.22) ±□
post	42.27 (7.14) ±□	46.00 (5.66) ±□
IRI: Interpersonal Reactivity Index		
IRI PD: Personal Distress		
pre	12.00 (6.88)	10.56 (2.60)
post	11.45 (5.57)	10.22 (3.15)

Table 4: Pre- and Post-Intervention Mindfulness and Empathy Subscales. Note: Pre: One week before intervention. Post: One week after intervention. Intervention: M&H Program (six sessions over three months).

± statistically significant time by group interaction effect p<0.05.
** statistically significant main effect of time p<0.05.*
□ statistically significant main effect of time p<0.01.

Analyses of outcomes (i.e., all instrument subscale and total scale scores) were conducted with mixed RM-ANOVA, using time of assessment (i.e., pre- and post-intervention) as a within-subjects factor and study group (i.e., intervention or control) as a between-subjects factor. One intervention-group participant's data for the EQ D subscale was excluded due to a missing data point; thus, analysis of EQ D for the intervention group was based on eight participants, instead of nine. Changes in pre- and post-intervention levels of mindfulness and empathy are depicted with 95% confidence intervals for the mean. See Figure 1 for FFMQ subscales, Figure 2 for EQ subscales and Figure 3 for IRI PD (empathy instrument) subscale.

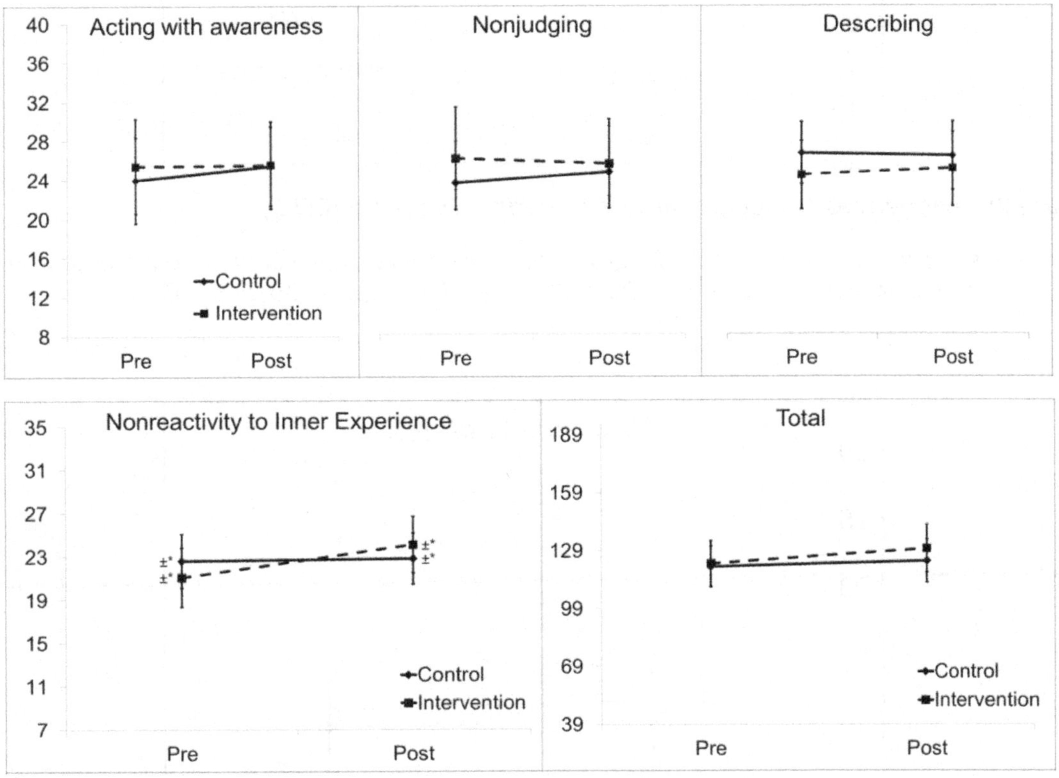

Figure 1: Five Facet Mindfulness Questionnaire (FFMQ). *Top panel: No significant main effect of time or time by group interaction effect for: Acting with Awareness (FFMQ AA), Nonjudging of Experience (FFMQ NJ), and Describing Experience (FFMQ D), were not statistically significant (top panel). Nonreactivity to Inner Experience subscale (FFMQ NR) had a statistically significant time by group interaction effect p<0.05 (bottom left). Results for FFMQ Total (bottom right) were not statistically significant. FFMQ NR (p<0.05, ηp2=0.20 large effect)*

± statistically significant time by group interaction effect (p<0.05, ηp2=0.20 large effect)
** statistically significant main effect of time (p<0.05, ηp2=0.20 large effect)*

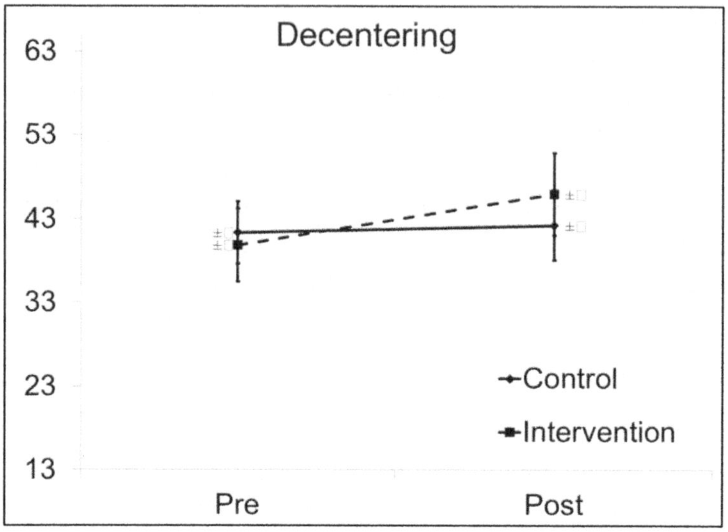

Figure 2: Experiences Questionnaire Decentering Subscale (EQ D).

± *statistically significant time by group interaction effect ($p<0.05$, $\eta p2=0.23$ large effect).*
□ *statistically significant main effect of time ($p<0.01$, $\eta p2=0.35$ large effect).*

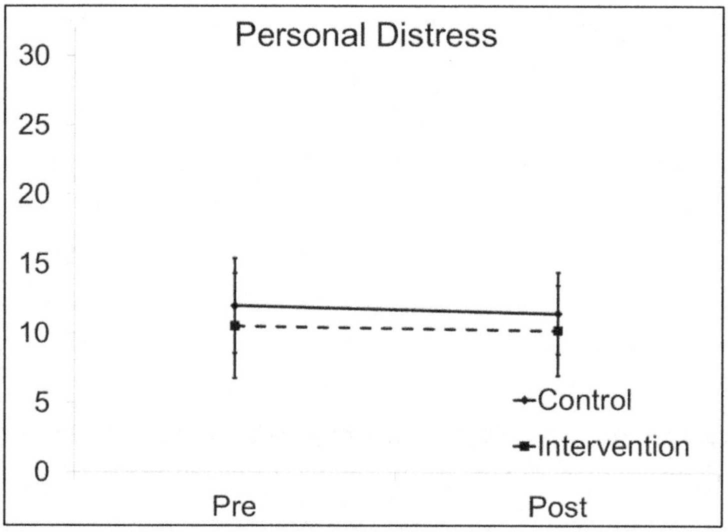

Figure 3: Interpersonal Reactivity Index Personal Distress subscale (IRI PD)*. Results were not statistically significant.*

Normality was assessed using Shapiro-Wilk's test and through visual examination of frequency distributions—all outcomes for both study groups were assessed to be drawn from a relatively normal population. Homogeneity of variance was examined using Levene's test—no violations of this assumption were observed. Effect sizes related to each mixed RM-ANOVA are reported as partial eta-squared (ηp2) values (see captions below Figures 1-3). P-values<0.05 were considered statistically significant. All quantitative analyses were conducted using SPSS (*SPSS* [computer program], 2014). Debrief Sessions were analyzed for themes using NVivo, a qualitative data analysis (QDA) computer program that allows researchers to keep track of how frequently topics appear in a text (*NVivo* [computer program], 2012). Each session was transcribed by one of two researchers and data was analyzed by the research team including the PI and the UCF COM Director of Faculty Development who has experience in medical education qualitative research. Nodes were created during group meetings based upon common themes. Some nodes were similar to themes measured by survey instruments, however, other nodes were unique to the debrief sessions. Grounded theory approach notes what themes organically arise in qualitative data and this is important to gain a broader understanding of data and suggest possible directions for future research. All nodes (i.e. themes) were defined by the research team and definitions, theme reference frequency, and session frequency are depicted in Figure 4.

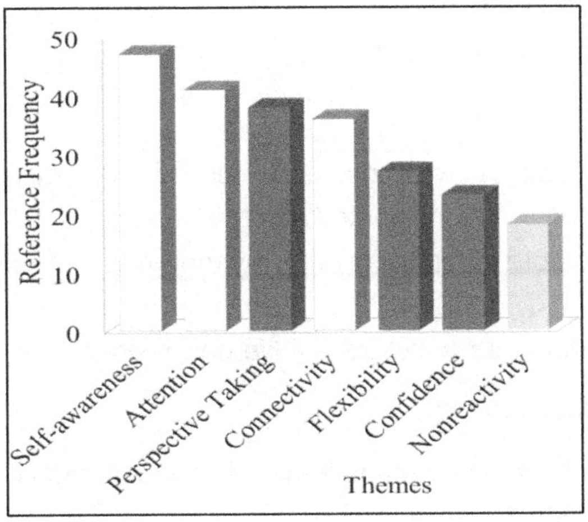

***Figure 4: Debrief Session Analysis**. Themes and Reference Frequency are presented from 6 debrief sessions. Colors indicate session frequency: White (6/6 sessions); Grey (5/6 sessions); Dark Grey (4/6 sessions). Theme Definitions: Self-awareness (Being aware of your feelings); Attention (Noticing things around you); Perspective Taking (Seeing things from another's point of view); Connectivity (Experiences where the participants are connected to other beings); Flexibility (Coming up with a new approach to a situation); Confidence (A feeling of trust or self-assurance arising from one's abilities or qualities); Nonreactivity (Being aware of your feelings and not reacting to them)*

RESULTS

Twenty (intervention, n=9; controls, n=11; 55% female) first-year medical students at UCF COM volunteered to participate in this study. Mean and standard deviation for intervention and control groups, pre- and post-intervention are presented in Table 4.

Debrief Sessions

Debrief session analysis is presented in Figure 4. Debrief sessions were analyzed for frequency of themes with respect to references and sessions. Overall, attention and self-awareness (both aspects of mindfulness) were the most frequent themes found in debrief session analysis.

Qualitative Survey

Qualitative survey analysis is presented in Table 5 including common themes and excerpts as quotes. Themes which emerged included: Self-awareness and Nonreactivity; Connectivity and Perspective Taking; Confidence and Flexibility; and Generalizability to the Medical Field.

Table 5: Qualitative Survey Analysis

Self-awareness and Nonreactivity

- Better able to think about the emotions I'm feeling and why I might be feeling them, instead of letting the emotions control me.
- Allowed me to be more aware of myself, my thoughts and how I portray myself to others.
- My mood was better and my outlook on life has improved.

Connectivity and Perspective Taking

- A lot nicer to people and let them talk more in conversations and am paying more attention to what they are saying.
- Feel more comfortable that I did before the Program. I've noticed it is easier for me to "read" body language.
- More confident in using my nonverbal skills and more perceptive of the thoughts/feelings of others.
- Made me more considerate.

Table 5: Qualitative Survey Analysis – cont'd.

Confidence and Flexibility

- Much more confident in trying things without first having to watch someone else do it a few times.
- Helped me approach an unknown situation with confidence and intention.
- More confident in my abilities.
- More mindful and assertive.
- Even when it seems really difficult, there usually is a way to solve it.

Generalizability to Medical Field

- Sometimes it is difficult for me to put myself in another person's shoes and see things from their perspective, and I am concerned that this will affect my ability to build rapport with some of my future patients.*
- Very frustrated and feel as though I am failing them as a health care provider.*
- Noticed I was more aware of patients' body language and nonverbal skills during COP.**
- Working with horses is still very different than working with humans, so I doubt I would use the same techniques I used with the horses on people.
- I am more confident in interacting with patients and am aware of how my energy level affects them.

Table 5. Qualitative Survey Analysis. Note: Intervention group (n=9) unanimously stated they would recommend the M&H Program to other medical students.

** Indicates response to question asking participants to reflect on how they felt before beginning M&H.*
*** Community of Practice (COP) is a UCF COM program where medical students enhance their clinical skills by interacting with real patients under the supervision of physicians in the community.*

Discussion

Participants who completed the M&H Program had significantly higher post-intervention mindfulness scores compared to controls. Intervention group scores were significantly higher in the mindfulness instrument subscales FFMQ NR (measuring nonreactivity) and EQ D (measuring decentering). Nonreactivity and decentering are both related to mindfulness and important skills to practice and develop throughout medical school. Attention, self-awareness, connectivity, and confidence were also found to be increased in the analysis of the qualitative survey. Using a mixed methods design, this research presents quantitative data supplemented with qualitative data and highlights previously unmeasured changes in medical students. Many themes appearing in debrief sessions overlapped with statements found in the qualitative survey and further support our findings that participation in M&H leads to increased levels of mindfulness.

Mindfulness

Self-awareness was the most referenced theme throughout the debrief session and appeared in 6/6 debrief sessions (see Figure 4). Although not as common as self-awareness, nonreactivity appeared in 5 out of 6 debrief sessions. Self-awareness was defined by researchers as "being aware of your feelings." Nonreactivity was defined by researchers as "being aware of your feelings and not reacting to them." Although self-awareness and nonreactivity were examined as separate nodes in our grounded theory approach to debrief sessions, they are very similar concepts that relate to mindfulness as one must first be self-aware of one's feelings (self-awareness) before one can take a normative stance as to whether they are good or bad (nonreactivity).

Intervention group participants expressed increased levels of nonreactivity in their qualitative survey answers. For instance, in the qualitative survey, one participant states: "I am better able to think about the emotions I'm feeling and why I might be feeling them, instead of letting the emotions control me." Qualitative data such as that found in the debrief sessions and qualitative survey support the statistically significant findings in the FFMQ NR mindfulness instrument.

Nonreactivity is an important skill for medical students to develop. As an aspect of mindfulness, nonreactivity protects against depression (Brown & Ryan, 2003) and allows students to manage their reactions to challenging circumstances occurring in the classroom or hospital setting. One statement on the FFMQ NR subscale highlights this concept: "In difficult situations, I can pause without immediately reacting." It is important to react quickly in emergency situations; however, it is equally important to react calmly. Nonreactivity applies to stressors and emergencies in the classroom and hospital settings alike.

Similar to nonreactivity, decentering is an important skill related to mindfulness. One statement on the decentering subscale highlights the importance of decentering for medical students. It states: "I am kinder to myself when things go wrong." Decentering is similar to self-compassion, a self-action with which many medical students struggle as they are motivated to continually succeed. The same drive that propels individuals for success in gaining admission to medical school may negatively affect their mental health, and, ironically, their success within medical school as challenges escalate. It is important for medical students to be both motivated and accepting of themselves and others in order to succeed academically, socially, and emotionally.

Empathy

We hypothesized that the M&H Program would benefit medical students by teaching empathy skills, however, we did not find significant increases in empathy as measured by the IRI Personal Distress subscale (Davis, 1983). It is possible that this is due to a ceiling effect, where medical students already have high levels of empathy. Unlike mindfulness, empathy is a quality that many medical schools select for in their applicants, and many individuals who apply to medical school may have higher baseline levels of empathy. It is also possible that changes in empathy were not found due to a small sample size of participants.

Despite the failure to find significant changes in empathy we can assess changes relating to empathy as presented in the qualitative data. After self-awareness and attention, the two most common themes found in debrief sessions were perspective taking and connectivity. Perspective taking was defined by researchers as "seeing things from another's point of view" and connectivity was defined as "experiences where the participants are connected to other beings."

We believe that participating in the M&H Program forces students to pause and be aware of the environment, themselves, and others. As one student notes in the qualitative survey: "I am a lot nicer to people and let them talk more in conversations and am paying more attention to what they are saying." Studies have found that patients are more likely to share pertinent information with physicians who display attentiveness and empathy (Halpern, 2007). Empathy skills are important for information-gathering in clinical scenarios and further research should examine whether M&H improves empathy.

CONFIDENCE AND FLEXIBILITY

Although we did not set out to measure confidence and flexibility we found through qualitative data that students self-reported increased levels in these items after completing the M&H Program. Many of the M&H Program objectives found in the M&H manual and listed in Table 1 were reflected in qualitative data collected in this study. Many intervention group participants reported increased levels of confidence after completing the M&H Program and throughout M&H Program debrief sessions. In addition, increases in confidence may be generalizable to the medical field. One student noted in the qualitative survey: "I noticed I was more aware of patients' body language and nonverbal skills during COP." COP, also known as community of practice, is a UCF COM program where medical students interact with patients in clinical settings under physician supervision.

It is important for medical schools to instill in students self-confidence as well as respect for others. Despite successful admission to medical school, many medical students struggle with confidence in the classroom—they may wish to practice but do not feel comfortable in situations such as volunteering to be the first student to practice a technique in a clinical setting. Conversely, other students might be eager to volunteer, but lack awareness of others and fail to encourage classmates which can negatively impact their interactions with peers, faculty, and patients. The M&H Program can benefit both types of students by encouraging those who are more reticent to take chances and broaden the awareness of all students so that they may better interact with their classmates and develop emotional intelligence. One M&H participant notes in the qualitative survey: "I am much more confident in trying things without first having to watch someone else do it a few times."

ALTERNATIVE EXPLANATIONS AND LIMITATIONS

Participation in the M&H Program is a novel experience, however, we believe changes in the intervention group were due to our intervention and not due to a novelty effect as we found high instrument reliability on the mindfulness instrument and further we surmise that the novelty effect would dissipate over time (by the end of six sessions). Another alternative explanation for our findings is that the intervention group participated in an enjoyable activity that brought them outside the classroom and it is a fun outdoor activity, and that it was this, and not participation in the M&H program, which improved intervention group mindfulness. Our qualitative data suggests that this is not the case.

An alternative explanation for failure to find changes in empathy is that the M&H program is not effective in changing empathy for other humans, possibly due to the use of animals. One student noted in the qualitative survey that interacting with horses is very different from interacting with patients, and that student had reservations about whether they would use the same skills with patients. At this time, it is not possible to come to a conclusion about possible changes in empathy since the instrument used in the study, the IRI, had low instrument reliability, and as a result, three out of the four subscales were excluded from final data analysis. We recommend further data collection measuring changes in empathy using a different empathy instrument.

This study has a small sample size, which leads our study to be underpowered. Although the small sample size limits power, this is an ongoing study, and in the future more data will be collected. Another limitation is that our participants were not randomly assigned. Analysis of baseline differences failed to reveal significant differences between intervention and controls in all survey instruments, which adds strength to our conclusions.

Results from this study indicate that medical students may benefit from an M&H program by learning mindfulness skills and building confidence through EAL. This study builds on previous research on M&H at UCF COM, which hypothesized that structured interactions with horses through EAL may be a useful intervention in medical education. An M&H program may supplement traditional educational approaches as a unique form of experiential learning. Our results are promising for further data collection, and our qualitative findings suggest possible directions for future research such as evaluating the effects of an M&H program on confidence.

REFERENCES

Ammentorp, J., Sabroe, S., Kofoed, P. E., & Mainz, J. (2007). The effect of training in communication skills on medical doctors' and nurses' self-efficacy. A randomized controlled trial. *Patient Education & Counseling*, 66 (3), 270-277.

Baer, R. A., Smith, G. T., Hopkins, J., Krietemeyer, J., & Toney, L. (2006). Using self-report assessment methods to explore facets of mindfulness. *Assessment*, 13 (1), 27-45.

Beach, M. C., Roter, D., Korthuis, P. T., Epstein, M. C., Shap, V., Ratanawongsa, N.,... Saha, S. (2013). A multicenter study of physician mindfulness and health care quality. *Annals of Family Medicine*, 11 (5), 421-428.

Beckman, H. B., Wendland, M., Mooney, C., Krasner, M. S., Quill, T. E., Suchman, A. L., & Epstein, R. M. (2012). The impact of a program in mindful communication on primary care physicians. *Academic Medicine*, 87 (6), 815-819.

Bishop, S. R., Lau, M., Shapiro, S., Carlson, L., Anderson, N. D., Carmody, J.,... Devins, G. (2004). Mindfulness: A proposed operational definition. *Clinical Psychology: Science and Practice*, 11 (3), 230-241.

Brown, K. W., & Ryan, R. M. (2003). The benefits of being present: mindfulness and its role in psychological well-being. Journal of Personality and Social Psychology, 84 (4), 822-848.

Davis, M. H. (1983). Measuring individual differences in empathy: Evidence for a multidimensional approach. *Journal of Personality and Social Psychology*, 44 (1), 113-126.

Fresco, D. M., Moore, M. T., van Dulmen, M. H., Segal, Z. V., Ma, S. H., Teasdale, J. D., & Williams, J.M. (2007). Initial psychometric properties of the experiences questionnaire: Validation of a self-report measure of decentering. *Behavior Therapy*, 38, 234-246.

Gonzalez, A. (2016, March). Medicine and horsemanship: Evaluation of the effect of equine-assisted learning and therapy on medical students' perceived levels of stress and depression. Paper presented at FIRE Research Conference, University of Central Florida College of Medicine, Orlando, FL.

Halpern, J. (2007). Empathy and patient-physician conflicts. *Journal of General Internal Medicine*, 22 (5), 696-700.

Kane, B. (2007). The Manual of medicine and horsemanship: Transforming the doctor-patient relationship with equine-assisted learning. Bloomington, IN: AuthorHouse.

Kostamo, K. (2015, March). Medicine and horsemanship: Evaluation of the effect of equine-assisted activities and therapy on medical student self-efficacy and nonverbal communication skills. Paper presented at FIRE Research Conference, University of Central Florida College of Medicine, Orlando, FL.

MacLean, B. (2011). Guest editorial: equine-assisted therapy. *Journal of Rehabilitation Research and Development,* 48 (7), ix-xii.

Masini, A. (2010). Equine-assisted psychotherapy in clinical practice. *Journal of Psychosocial Nursing and Mental Health Services*, 48 (10), 30-34.

Merriam, S. (2009). Qualitative Research: A guide to design and implementation. San Francisco, CA: Jossey-Bass.

Nevins, R., Finch, S., Hickling, E. J., & Barnett, S. D. (2013). The Saratoga WarHorse project: a case study of the treatment of psychological distress in a veteran of Operation Iraqi Freedom. *Advances in Mind-Body Medicine*, 27 (4), 22-25.

NVivo [computer program]. Version 10. (2012). Melbourne, Australia: QSR International.

Park, T., Reilly-Spong, M., & Gross, C. R. (2013). Mindfulness: a systematic review of instruments to measure an emergent patient-reported outcome (PRO). *Quality of Life Research*, 22 (10), 2639-2659.

Patel, P. (2015, March). Medicine and horsemanship: Evaluation of the effect of equine assisted activities and therapy on medical students' depression levels and stress relief. Paper presented at FIRE Research Conference, University of Central Florida College of Medicine, Orlando, FL.

Priesthoff, A. (2015, March). Medicine and horsemanship: Evaluation of the effects of equine assisted activity and therapy on medical students' mindfulness and social self-efficacy skills. Paper presented at FIRE Research Conference, University of Central Florida College of Medicine, Orlando, FL.

Qualtrics [computer program]. Version 6209708. (2015). Provo, Utah: Qualtrics.

Rasmussen, M. K., & Pidgeon, A. M. (2011). The direct and indirect benefits of dispositional mindfulness on self-esteem and social anxiety. *Anxiety, Stress & Coping*, 24 (2), 227-233.

Sigsbee, B., & Bernat, J. L. (2014). Physician burnout: A neurologic crisis. *Neurology*, 83 (24), 2302-2306.

SPSS [computer program]. Version 22. (2014). Armonk, NY: IBM.

Vaizer, J. (2014, March). Medicine and horsemanship: Evaluation of the effect of equine-assisted learning and therapy on medical students' perceived levels of stress and depression. Paper presented at FIRE Research Conference, University of Central Florida College of Medicine, Orlando, FL.

Appendix A: Five Facet Mindfulness Questionnaire (FFMQ)

1. When I'm walking, I deliberately notice the sensations of my body moving. (OBS)
2. I'm good at finding words to describe my feelings. (D)
3. I criticize myself for having irrational or inappropriate emotions. (NJ-R)
4. I perceive my feelings and emotions without having to react to them. (NR)
5. When I do things, my mind wanders off and I'm easily distracted. (AA-R)
6. When I take a shower or bath, I stay alert to the sensations of water on my body. (OBS)
7. I can easily put my beliefs, opinions, and expectations into words. (D)
8. I don't pay attention to what I'm doing because I'm day dreaming, worrying, or otherwise distracted. (AA-R)
9. I watch my feelings without getting lost in them. (NR)
10. I tell myself I shouldn't be feeling the way I'm feeling. (NJ-R)
11. I notice how foods and drinks affect my thoughts, bodily sensations, and emotions. (OBS)
12. It's hard for me to find the words to describe what I'm thinking. (D-R)
13. I am easily distracted. (AA-R)
14. I believe some of my thoughts are abnormal or bad and I shouldn't think that way. (NJ-R)
15. I pay attention to sensations, such as the wind in my hair or sun on my face. (OBS)
16. I have trouble thinking of the right words to express how I feel about things. (D-R)
17. I make judgments about whether my thoughts are good or bad. (NJ-R)
18. I find it difficult to stay focused on what's happening in the present. (AA-R)
19. When I have distressing thoughts or images, I "step back" and am aware of the thought or image without getting taken over by it. (NR)
20. I pay attention to sounds, such as clocks ticking, birds chirping, or cars passing. (OBS)
21. In difficult situations, I can pause without immediately reacting. (NR)
22. When I have a sensation in my body, it's difficult for me to describe it because I can't find the right words. (D-R)
23. It seems I am "running on automatic" without much awareness of what I'm doing. (AA-R)
24. When I have distressing thoughts or images, I feel calm soon after. (NR)
25. I tell myself that I shouldn't be thinking the way I'm thinking. (NJ-R)
26. I notice smells and aromas of things. (OBS)
27. Even when I'm feeling terribly upset, I can find a way to put it into words. (D)
28. I rush through activities without being really attentive to them. (AA-R)
29. When I have distressing thoughts or images I am able just to notice them without reacting. (NR)
30. I think some of my emotions are bad or inappropriate and I shouldn't feel them. (NJ-R)
31. I notice visual elements in art or nature, such as colors, shapes, textures, or patterns of light and shadow. (OBS)
32. My natural tendency is to put my experiences into words. (D)
33. When I have distressing thoughts or images, I just notice them and let them go. (NR)
34. I do jobs or tasks automatically without being aware of what I'm doing. (AA-R)

APPENDIX A: FIVE FACET MINDFULNESS QUESTIONNAIRE (FFMQ) – CONT'D.

35. When I have distressing thoughts or images, I judge myself as good or bad, depending what the thought/image is about. (NJ-R)
36. I pay attention to how my emotions affect my thoughts and behavior. (OBS)
37. I can usually describe how I feel at the moment in considerable detail. (D)
38. I find myself doing things without paying attention. (AA-R)
39. I disapprove of myself when I have irrational ideas. (NJ-R)

Note: R=reverse-scored; FFMQ Observing of Inner Experience (OBS; 8 Items): 1, 6, 11, 15, 20, 26, 31, 36; Describing Experience (D; 8 Items): 2, 7, 12R, 16R, 22R, 27, 32, 37; Acting with Awareness (AA; 8 Items): 5R, 8R, 13R, 18R, 23R, 28R, 34R, 38R; Nonjudging of Experience (NJ; 8 Items): 3R, 10R, 14R, 17R, 25R, 30R, 35R, 39R; Nonreactivity to Inner Experience (NR; 7 Items): 4, 9, 19, 21, 24, 29, 33; FFMQ Total: sum of subscales.

APPENDIX B: EXPERIENCES QUESTIONNAIRE (EQ)

1. I think about what will happen in the future. (R)
2. I remind myself that thoughts aren't facts. (removed from analysis)
3. I am better able to accept myself as I am.(D)
4. I notice all sorts of little things and details in the world around me. (R)
5. I am kinder to myself when things go wrong. (D)
6. I can slow my thinking at times of stress. (D)
7. I wonder what kind of person I really am. (R)
8. I am not so easily carried away by my thoughts and feelings. (D)
9. I notice that I don't take difficulties so personally. (D)
10. I can separate myself from my thoughts and feelings. (D)
11. I analyze why things turn out the way they do. (R)
12. I can take time to respond to difficulties. (D)
13. I think over and over again about what others have said to me. (R)
14. I can treat myself kindly. (D)
15. I can observe unpleasant feelings without being drawn into them. (D)
16. I have the sense that I am fully aware of what is going on around me and inside me. (D)
17. I can actually see that I am not my thoughts. (D)
18. I am consciously aware of a sense of my body as a whole. (D)
19. I think about the ways in which I am different from other people. (R)
20. I view things from a wider perspective. (D and R)

APPENDIX C: INTERPERSONAL REACTIVITY INDEX (IRI)

1. I daydream and fantasize, with some regularity, about things that might happen to me. (FS)
2. I often have tender, concerned feelings for people less fortunate than me. (EC)
3. I sometimes find it difficult to see things from the "other guy's" point of view. (PT-R)
4. Sometimes I don't feel very sorry for other people when they are having problems. (EC-R)
5. I really get involved with the feelings of the characters in a novel. (FS)
6. In emergency situations, I feel apprehensive and ill-at-ease. (PD)
7. I am usually objective when I watch a movie or play, and I don't often get completely caught up in it. (FS-R)
8. I try to look at everybody's side of a disagreement before I make a decision. (PT)
9. When I see someone being taken advantage of, I feel kind of protective towards them. (EC)
10. I sometimes feel helpless when I am in the middle of a very emotional situation. (PD)
11. I sometimes try to understand my friends better by imagining how things look from their perspective. (PT)
12. Becoming extremely involved in a good book or movie is somewhat rare for me. (FS-R)
13. When I see someone get hurt, I tend to remain calm. (PD-R)
14. Other people's misfortunes do not usually disturb me a great deal. (EC-R)
15. If I'm sure I'm right about something, I don't waste much time listening to other people's arguments. (PT-R)
16. After seeing a play or movie, I have felt as though I were one of the characters. (FS)
17. Being in a tense emotional situation scares me. (PD)
18. When I see someone being treated unfairly, I sometimes don't feel very much pity for them. (EC-R)
19. I am usually pretty effective in dealing with emergencies. (PD-R)
20. I am often quite touched by things that I see happen. (EC)
21. I believe that there are two sides to every question and try to look at them both. (PT)
22. I would describe myself as a pretty soft-hearted person. (EC)
23. When I watch a good movie, I can very easily put myself in the place of a leading character. (FS)
24. I tend to lose control during emergencies. (PD)
25. When I'm upset at someone, I usually try to "put myself in his shoes" for a while. (PT)
26. When I am reading an interesting story or novel, I imagine how I would feel if the events in the story were happening to me. (FS)
27. When I see someone who badly needs help in an emergency, I go to pieces. (PD)
28. Before criticizing somebody, I try to imagine how I would feel if I were in their place. (PT)

Appendix D: M&H Program Session Descriptions

Session One

- Safety demonstration: One horse at liberty in a corral. Lead rope optional. The facilitator gets into the corral with the horse while students remain outside. Talk includes horse body language such as ear positions, tail swishing, mouthing behaviors, and swinging the haunches around.
- Round up/Turn out: Group observation of at least 3-5 horses newly introduced into a space together. From outside the corral observe the body language of horses and horse hierarchies.
- Con Su Permiso - Approach the Horse: 3-5 horses in a corral - goal is to notice when the horse notices you. Students approach the horses and announce "now" when the horse notices them.

Session Two

- Choose or be Chosen: One horse per 3-4 students at liberty in a large arena. Students are in the arena interacting with the horses however they wish.
- Grooming: One horse tied for every 1-4 students with grooming boxes. Students are taught how to use the grooming tools. Students are advised that horses differ widely in how they respond to grooming, and that some have ticklish areas that they might object to having touched. Optional: teach students equine massage techniques.
- Hind Leg Release: It is desirable to mix easy lifters with recalcitrant lifters and horses with long and slow releases so that students feel the difference. Facilitators should demonstrate the farrier's hold of the hind leg, and narrate the pull-in and release. The effect can be felt without using the farrier's position, but proper alignment with the horses' body and support of the fetlock joint should be demonstrated.

Session Three

- The Four Ways of Wisdom: Four horses cross-tied. Three students per horse. Ask the students to describe the horses' outer characteristics (Ex: brown, long mane, wearing shoes). Ask the students to describe their sensory experience of the horse (Ex: warm, soft, musty smelling). Ask the students to describe how they feel about the horse (Ex: afraid, loving, amused). Ask the students to close their eyes and ask the horse a question and imagine the answer
- Vet Exam: 4 horses cross-tied. 3 students per horse. Assess capillary refill in the gums, jugular venous filling, skin tenting, cecal and colonic gut sounds, muscle tone, back soreness, and heart sounds

Session Four

- Am I being Herd?: One horse per two students in an arena, initially with halters and lead ropes. (Five horses) Half of the group goes first. Each person leads their horse by lead rope in a group circle in both clockwise and counterclockwise directions. Remove the halters and lead ropes and repeat the clockwise and counterclockwise directions. Repeat the activity with the other half of the group (switch roles).
- Uniformed consent: Get a horse to walk over a tarp with scary obstacles.
- Individual longeing: One person longeing one horse at a time with and without a longe line in a round pen.

Session Five

- Pas de Dirt: One horse with a halter and lead line for every 1-4 students in the arena. Each student is asked to trot the horse out and back 30 feet on a lead line. Then they have the horse trot in a 20-foot circle clockwise and counterclockwise on the line. Option: Students may try different horses to experience different levels of difficulty.
- Horse Pool: Set-up 6 "U-shaped" pockets (using three ground poles) around the arena to resemble a pool table. Students in groups of 3-4 must get their horse to stand in each of the pockets for a few seconds, without touching or bribing the horse.
  Note: This activity was developed by Principal Investigator and is not found in M&H Handbook

Session Six

- Rotating Extended Appendages: One or two horses in the arena. A halter with lead rope is buckled onto the railing of the arena. A saddle or (or bareback pad) with blanket is over the fence of the arena. A second horse may be used to offer a distraction to the horse being tacked up and to elicit a decision process about which horse to catch. Option: There is a book on how to tack up a horse. Four students line up and link elbows, and they must successfully halter and saddle a horse of their choice. The two students on the ends of the line can use their right and left arm; they can act, but they cannot speak. The two students in the middle are the "brains" that tell the arms step-by-step what to do; they can speak, but they cannot act. If no rule violations happen in three minutes, the line may rotate positions. Option: Outside "Chief Resident" keeps track of violations.

ARTICLE TWO

Conversational Skills Instruction During Therapeutic Horseback Riding for Adolescents with Mild Disabilities

Laura A. Bassette, Ph.D., BCBA-D, Department of Special Education, Ball State University
Teresa Taber-Doughty, Ph.D., Dean, College of Education, University of Texas at Arlington
Colleen Brady, Ph.D., Department of Youth Development and Agricultural Education, Purdue University

Correspondence to:
Dr. Laura Bassette
Department of Special Education, Teachers College, Room 705
Ball State University, Muncie, IN 47306
Email: labassette@bsu.edu

CONVERSATIONAL SKILLS INSTRUCTION DURING THERAPEUTIC HORSEBACK RIDING FOR ADOLESCENTS WITH MILD DISABILITIES

ABSTRACT

This study examined the effects of using direct instruction and verbal prompting to teach conversational skills to adolescents with mild disabilities during therapeutic horseback riding (TR) lessons. A multiple probe design was used to assess the participants' fluency and generalization of conversational skills practiced during riding lessons. Participants received direct instruction and were prompted to practice conversational skills with their horses, other riders, volunteers, and the TR instructor during lessons. Following the TR lessons, participants completed scavenger hunts during un-mounted activities to assess their fluency and generalization of skills. The results indicate all three participants demonstrated increases in fluency of skills and two participants maintained the skills one month following intervention. Suggestions for future research and implications for practices are discussed.

KEYWORDS

Therapeutic Riding; Conversational Skills; Mild Disabilities.

INTRODUCTION

Adolescents with disabilities served in classrooms with non-disabled peers may lack social skills, which hinder their ability to engage with classmates (Smith & Tyler, 2010). Specifically, an adolescent with mild disabilities such as attention deficit hyperactivity disorder (ADHD) and high functioning autism (HFA) may be socially expected to engage appropriately, yet have difficulty doing so. The DSM-5 criterion includes impediments in social interactions for children with ADHD and autism (American Psychiatric Association, 2013) and these children may experience a lack of awareness in others, difficulties in forming friendships, and a lack of empathy (Clark, Feehan, Tinline, & Vostanis, 1999). They may also have difficulty attending to others during conversations (McAffee, 2002), which could lead to isolation, difficulty in making friends, and potential bullying (Koenig, 2012; Kowalski & Fedina, 2011). Additionally, children with these disabilities may develop anxiety and depression (Sterling, Dawson, Estes, & Greenson, 2008).

Social skills' training (SST) is one intervention previously used for children with mild disabilities (Rao, Beidel, & Murray, 2008) to address these deficits. SST training was previously used to teach general conversational skills (Tse et al., 2007), initiating conversations (Sansosti & Powell-Smith, 2006), and greeting people (Barry et. al., 2003). Despite this research, Bellini (2008) suggests previous research in SST has not always matched the intervention to the deficits. Bellini suggests interventions target specific deficits, for instance, an acquisition deficit intervention would focus on teaching and practicing new skills while a performance deficit intervention would seek to improve performance of known skills (focus on fluency and generalization). The effectiveness of SST is mixed and additional research is needed (de Boo & Prins, 2006).

One way to enhance social skills training for children with disabilities may be through the incorporation of animals. Previous research found petting animals decreases anxiety (Shiloh, Sorek, & Terkel, 2003) and contact with animals decreased the heart rate in children during stressful tasks (Friedmann et al., 1983). Additionally, Rud & Beck (2000) noted a calming effect on students in a special education classroom who held classroom pets. Specific research in equine assisted activities (EAA) has provided support regarding the impact of horses on social functioning in children with autism (Bass, Duchowny, & Llabre, 2009) and found children who received therapeutic riding (TR) had significant increases in responsiveness and motivation. Additional research found EAA such as TR impacted empathy (Anderson & Meints, 2016) and non-riding activities positively impacted academic engagement (Bassette et al., 2014).

Taken together, these studies demonstrate emerging evidence of EAA as a means to increase social functioning; however, there remains a need to focus on specific skills. Less empirical research was conducted examining the impact of EAA on students with ADHD; also, in 2015 the

Professional Association for Therapeutic Horsemanship International (PATH Intl.) reported 664 of 877 PATH centers served children with ADHD indicating a belief that there are benefits (PATH Intl., 2015).

Adolescents with mild disabilities require a specific skill set that allows them to interact with others effectively and fluently. Furthermore, there is a need for children with disabilities to generalize previously acquired skills to new settings and people (Rao et al., 2008). The previous research suggests teaching conversational skills while riding may allow students to be less anxious while practicing skills. This decrease in anxiety may allow students with disabilities to increase acquisition, performance, and fluency and increase generalization to conversations with other people.

There is also a need for practitioners of TR to implement and use evidenced-based practices (EBP) (Fine, 2006). Direct instruction and verbal prompting are considered educational EBPs (Smith & Tyler, 2010) and are frequently implemented into TR settings; yet, the effectiveness of these strategies within TR was not previously examined. The purpose of the present study was to examine if TR instructors could utilize direct instruction and verbal prompting to teach conversational skills to children with mild disabilities during TR and explore if the participants could then generalize the targeted skills during non-riding activities.

METHOD

PARTICIPANTS

Institutional Review Board approval was received prior to recruiting participants and participants were recruited through a convenience sample. Three adolescent males, named for the purpose of the study as Walter, Adam, and Justin participated. Each was selected through a purposive sample based on willingness to participate, age, cognitive level, language ability, and TR experience. Inclusion criteria included: age (adolescence), disability that impaired social interactions, communicative speech, and acquisition but lack of fluency in conversational skills (Table 1).

Walter is a 14-year old Caucasian male diagnosed with Attention Deficit Hyperactivity Disorder (ADHD) and a secondary anxiety disorder. Walter was home schooled for two years prior to the study but attended public school in 2nd-6th grade. He knew basic conversational skills, however his mother indicated he struggled when verbally communicating with peers. She also indicated

Walter experienced some difficulty when interacting with his parents and could become defiant at times, but was overall compliant with adults other than his parents.

Adam is a 13-year old Caucasian male diagnosed with high functioning autism (HFA) and a secondary learning disability. Adam attended private school and was included in general education classes. At the start of the study, he read at a 7th grade level. Despite strong verbal skills, Adam would only initiate conversations when interested in the topic; otherwise he would not speak unless directly addressed. Adam previously attended a social group for adolescents with HFA where he worked on initiating conversations and asking questions to his peers. He was cooperative with adults, but was very quiet and nervous during interactions.

The third participant, Justin is a 13-year old Caucasian male diagnosed with HFA and was in 7th grade. Justin attended public school until 2nd grade and was then home schooled. In regards to reading and writing ability, he functioned on grade level and had strong verbal skills. Justin knew how to introduce himself and respond to others who initiated conversations with him, but demonstrated difficulty with listening skills and appropriately initiating conversations with others. He attended a social skills class one day a week where he worked on asking for help, relaxing during conversations, awareness of tone of voice, and listening to others while they are talking. When frustrated or anxious, Justin could become argumentative with adults and peers.

Settings and Materials

The study took place at a PATH (previously NARHA) Premier Accredited therapeutic horseback-riding center in the United States during a Summer Adventures Camp. All instructors were PATH certified. The camp was a five-day camp with programming each day from 9 am until noon. During camp, riders engaged in three hour-long activities including a mounted TR lesson, an unmounted barn activity, and an indoor classroom horse-related craft activity.

All phases of the study took place at the riding center. The main building of the center included an indoor riding arena adjacent to the barn area with stalls, a tack room, a classroom, and an observation area. The facility also included an outdoor arena connected to the indoor arena by a gate and outdoor riding trails. The intervention phase took place in the arenas and trails where riders received their TR instruction. During intervention, all riders participated in the same TR lesson and each rode their horse for the duration of each intervention session. Rider and horse teams also included a trained horse handler to help ensure safety.

The assessments for the three phases (baseline, intervention, and maintenance) took place throughout the main building including the indoor arena, barn area, tack room, and classroom. During assessment, participants were instructed to complete a scavenger hunt and gather items hidden in the barn area (classroom, aisle way, tack room, arena) and on foot (without their horses).

Horses. During the TR lessons, each rider was assigned a horse to ride. Each horse selected into the TR program was trained and tested for safety by the TR center. Assigned tack and equipment was used during the riding activities. Additionally, all horse and rider teams had a trained horse handler assigned to work with them and all riders wore approved riding helmets.

Scavenger Hunt Lists. Each list included a heading at the top center of the sheet and four pictures along with the written words describing the item (e.g., Appendix A). .

Experimental Design

A multiple probe design was used to determine the effectiveness of the intervention on targeted skills. The multiple probe design was selected as it decreases the collection of data across the multiple baselines while ensuring no changes in behavior prior to the implementation of the intervention. The multiple probe design also allowed for the determination of a functional relationship between the intervention (direct instruction and verbal prompting during TR lessons) and the rider's conversational skills as measured during the skill assessments (Kennedy, 2005). The design was also selected as it is commonly used when learning is involved (Alberto & Troutman, 2006).

Data Collection

Event recording was used to determine the percentage of targeted skills exhibited and scavenger hunts sessions were conducted 1:1 with the researcher. Participants completed a four-item scavenger hunt list during each assessment. For each of the four items, participants could potentially engage in five targeted skills for a total of 20. Each participant was directly observed during the scavenger hunt activity and the percentage of skills demonstrated out of 20 was calculated for each session.

Procedure

Baseline. The baseline condition was staggered across participants based on the order they were selected to begin intervention. Intervention activities (direct instruction and verbal prompting on targeted conversational skills) were not implemented during baseline. During baseline, each participant was given a scavenger hunt list with four items on it and told they needed to ask for the items before taking it. The researcher recorded if participants demonstrated any of the five-targeted conversational skills for each of the items on the list during baseline.

Intervention. During intervention, participants received direct instruction and were verbally prompted to engage in the targeted verbal conversational skills. Participants practiced the skills both with their horses as well as with people during the TR lesson. They were only prompted to engage in the targeted social skills during the TR lessons and were not prompted to engage in the skills during other camp activities (i.e., barn and craft sessions). Participants were taught three sets of targeted skills. For initiating a conversation, participants were taught to say "Hey" or "Excuse me" (component 1) and say the person or horse's name (component 2). Second, they were instructed to say "Please" (component 3) and phrase statements in the form of a question by using "May", "Will", "Would", "Can", or "Could" (component 4). Finally, they were instructed on how to end a conversation by saying a form of "Thanks" (component 5) once they received the item they requested.

The TR instructor provided direct instruction and verbal prompting to each participant during the mounted lessons. The horse handlers also assisted with prompting participants to engage in the targeted skills. Following direct instruction on how, when, and why participants should use these skills, they were verbally prompted to practice the targeted skills. They would do so by talking to their horses. For example, "Hey______(horse's name), will you please _______(walk on, whoa, turn left, etc.)?" When the horse complied, the participant would be taught to say, "Thank you for walking on." Participants also practiced these skills with one another during the riding lessons. For example, the instructor would prompt participants to ask one another for feedback. For example, one would say, "Excuse me _______ (person's name), can you please watch me do the _____ (posting trot, two-point, etc.)?" Then the participant completed the tasks and asked "How did I do" of the second participant, the second participant then provided feedback and the first participant said "Thanks"

Following intervention, participants were given the 4 item scavenger hunt list to assess their use of the targeted skills and prompted to collect the items using the same procedures as in baseline.

Maintenance. One month following intervention activities, participants completed an assessment in which the maintenance of targeted communication skills was analyzed using baseline procedures (i.e., no direct instruction or prompting provided). Specifically, participants were given a four item scavenger hunt list and instructed to find the items and ask for the items. The percent of targeted skills used by each participant out of 20 was recorded.

Social Validity

Participants and TR instructors completed two open-ended questionnaires. The first was completed prior to the start of the study on the first day of the camp and the second following the completion of the study on the last day of camp. Both participants and instructors were asked about likes/dislikes of teaching and learning conversational skills during the TR lessons. Initially, each participant indicated they enjoyed riding the horses and all reported riding was a relaxing activity they enjoyed. Participants also indicated they knew conversational skills were important and Walter indicated he thought communicating with the horses could help him communicate with people. The TR instructor indicated she thought focusing on the skills during lessons might help her be a better instructor and indicated these skills are important because they can be used in everyday life. Additionally, the instructor noted these skills might help improve the participants' confidence when communicating with others.

At the completion of the study, each participant indicated he felt more confident in his conversational skills after riding and the intervention helped them learn. Walter indicated he enjoyed practicing being polite and felt the scavenger hunts were fun. Adam indicated practicing these skills while riding helped him know what to say. He also expressed that he liked practicing with the horses and that it helped him communicate more effectively. The TR instructor stated that she liked having the participants' focus on conversational skills and that she believed it helped them to increase their confidence and ability to work with one another. She also indicated this was better compared to her traditional TR lessons because it involved the riders in the process. She said watching them encourage one another was helpful because the skills they were learning could be used in other areas of their lives.

Reliability

A second trained observer collected interobserver reliability data for each participant. The second observer recorded the targeted skills exhibited during the scavenger hunts simultaneously and independently from the primary researcher. Interobserver agreement data were calculated by dividing the number of agreements by the number of agreements plus disagreements and then multiplying by 100. For Walter and Adam, reliability data were collected for 40% of the baseline sessions, 43% of the intervention sessions, and 50% of the maintenance sessions. For Justin, reliability data were collected for 33% of the baseline sessions, 43% of the intervention sessions, and 50% of the maintenance sessions. Results ranged from 87.5-100% during baseline, 90-100% during intervention, and were 100% for all participants during maintenance.

Treatment Integrity

A checklist was created to ensure the TR instructors implemented the intervention appropriately. The checklist included three steps to ensure the instructor was teaching the targeted skills, allowing participants to practice those skills during the TR lessons, and instructing the volunteers to verbally prompt riders to engage in the skills. For all three riders, treatment integrity data were collected for 100% of the sessions and the average level of implementation was 91%. The treatment integrity was lower because the TR instructor did not review the target skills with any of the participants on the first day but this was corrected during subsequent sessions.

Results

Walter and Adam participated in four days of intervention activities while Justin participated in three days of intervention activities. Multiple assessments were conducted each day during the five-day camp. The results indicate participants had moderate increases in the percentage of targeted conversational skills used accurately. Each participant made relevant gains in performance between baseline and intervention and baseline and maintenance conditions (Table 1).

Table 1: Participant Information

Name	Age/Grade	IQ	Disability	Schooling	Prior TR experience
Walter	14, 9th grade	n/a	ADHD, anxiety disorder	Home	3 months
Adam	13, 8th grade	87	HFA	Private	5 years
Justin	13, 7th grade	74	HFA	Home	6 years

Walter

During baseline, Walter engaged in the targeted conversational skills 1% of the time. When visually analyzing data, an immediate increase can be seen during the initial three sessions of intervention and this remained higher than baseline with a mean of 70.7%. Additionally, the percentage of non-overlapping data between baseline and intervention was 100% indicating the intervention was highly effective for Walter. Variability was observed during intervention, however there was stability during maintenance. The mean during maintenance was 80% and was higher than both the baseline and intervention means. Additionally, the percent of non-overlapping data between baseline and maintenance was also 100%. These results indicate the intervention was highly effective for Walter.

Adam

During baseline, the mean targeted conversational skills were exhibited an average of 9%. This increased to 48.1% during intervention illustrating an average increase of 38% and these results were maintained at 45% during maintenance. Visual analysis show Adam had an immediate increase in the targeted conversational skills during the first session of intervention. While his results were variable during intervention, they were stable during maintenance. There was a sharp increase from the last session of baseline (20%) to the first session of intervention (80%) indicating an immediate change in level when the intervention began. In addition, a slight decrease in means was observed between intervention (47%) and maintenance (45%). The percentage of non-overlapping data between baseline and intervention was 86% and between baseline and maintenance it was 100% indicating the intervention was moderately effective for Adam.

Justin

The mean percent of skills exhibited by Justin during baseline was 13.3%. This increased to 32.9% during intervention and decreased to 18% during maintenance indicating the results were not maintained over time. Visual analysis indicates Justin had limited increases between baseline and intervention and data variability was evident during intervention. There was an immediate increase between the last session of baseline (15%) and the first session of intervention (55%) however the results were not maintained during the subsequent sessions of the intervention or during maintenance. The percent of non-overlapping data between baseline and intervention was 86%, however between baseline and maintenance the percent of non-overlapping data was 0%. These results indicate the intervention initially was moderately effective but not maintained over time.

Discussion

The purpose of this study was to examine the effect of using direct instruction and verbal prompting during TR lessons on generalization of conversational skills in adolescents with mild disabilities. The results showed an increase in use of the targeted conversational skills with some variability. All three experienced an immediate increase in their use of the skills at the start of intervention. The moderate gains observed in this study support previous research suggesting horses can help increase social interactions skills in children with disabilities (Bass, Duchowny, & Llabre, 2009).

All three participants indicated at the beginning of the study they found riding to be fun and relaxing. Previous research in social skills training suggests skills are more likely to be generalized when taught in these types of atmospheres (Gresham et al. 2001). While each enjoyed riding, there was variability in their performance, assessment of which took place during an unmounted activity. When practicing the skills during the TR lesson, participants were very willing to engage in skills with little prompting. However, they were not as willing to utilize the skills once asked to generalize the skills and the riding component was removed. It may be that the horse provided motivation and once removed from the setting, the participants were less motivated, or perhaps the horse allowed participants to be more relaxed physiologically. Physiology and anxiety data (e.g., blood pressure, cortisol levels) were not collected but it could be likely that the participants felt more anxious without the horse, which impacted their performance during the assessments. Previous research of animal-assisted activities has typically examined and found increases in social interactions in the participants when live animals are in the same setting as participants (Martin & Farnum, 2002; Sams, Fortney, & Willenbring, 2006). Future research should examine how the presence or absence of animals influences motivation and anxiety for children with disabilities when engaging in social interactions with both animals and people.

Walter had the greatest increase in performance and fluency of the targeted skills during intervention and maintenance. Walter enjoyed practicing the conversational skills with his horse while riding and was able to generalize the skills learned while riding to the assessments. He indicated he enjoyed the scavenger hunts, which may be part of the reason for his results. In addition, he did not appear to be anxious during the camp thus he might have felt more comfortable overall during the assessments.

Adam showed increases initially in performance during intervention; however, he lacked maintenance of the skills during the latter part of the intervention. He was willing to practice the skills with his horse during the riding lessons but had some difficulty in generalizing the skills to the assessment activity. He also had some hesitation and anxiety when talking to adults, which may have impacted his ability to engage in the targeted skills during the assessments. During session 12, Adam's performance dropped to 20%, which may be attributed to him not feeling well, however, Adam retained skills during maintenance. Additionally, following maintenance assessments, participants received a brief riding lesson. During this riding lesson, Adam spoke to his horse using the targeted conversational saying "Cowboy (horse's name), will you please walk on" unprompted. This occurred more than a month after the intervention indicating there was some maintenance of the skills learned during the intervention when placed in a TR setting.

Justin had the most variability in his results. There are several possible reasons for this. He only received three days of intervention due to the limited timeframe of the camp, while Walter and Adam received four days. Additionally, he reported experiencing high levels of anxiety due to a lack of sleep during camp. Justin did have initial increases in performance during intervention, however he had difficulty in generalizing the skills to the assessments and in practicing the skills with his horse during the TR lesson. He had participated in TR lessons at the center previously and was used to verbally directing the horse by saying only "Walk on" or using his legs and reins. When asked to practice the targeted conversational skills, he engaged in his previous skills instead of performing the targeted skills. Finally, Justin typically demonstrated resistance to change and his opposition to the intervention may possibly be attributed to asking him to change his behavior during intervention.

The results of this study have a number of significant implications regarding the implementation of conversational skills training in EAA. Specifically, the TR instructor was able to effectively implement with a high degree of fidelity conversational skills training during traditional TR lessons. All participants demonstrated increases in the performance (use) of skills; however, fluency and generalization were limited.

Children with mild disabilities such as ADHD and HFA are typically placed in inclusive settings with non-disabled peers and need to use basic conversational skills to have successful interactions with peers and teachers. Practitioners in the field of EAA can use similar evidence-based practices used by teachers to work on similar goals with their clients. It is important for practitioners to be knowledgeable of the teaching strategies used with children with disabilities to create fun and motivating ways to use these strategies when providing equine-assisted activities. This study expands on the previous literature in EAA and other animal based interventions (Martin & Farnum, 2002; Sams, Fortney, & Willenbring, 2006).

Limitations and Future Research

This was a pilot study with a number of limitations to be addressed in future research. First was the short time frame of this study. The entire study was completed during a 5-day period and the intervention was not equally implemented across all participants due to the time constraints of the TR camp. Future studies should take place over a longer time period and include a longer and more detailed intervention phase when teaching conversational skills. Additionally, participants included an adolescent with ADHD and adolescents with HFA and the total sample size was only three participants. There was no control group, which further limits the generalization of the results of this study. Future research may want to examine the effects of similar strategies using a larger sample, a control group, and randomized assignment so that results can be generalized.

Prior to the study, all of the participants were previously knowledgeable, having been taught the skills selected for the study. As illustrated in baseline, however, participants lacked performance, fluency, and generalizations of skills. Future research may want to examine teaching new acquisition of specific skills to children during equine activities as well as the impact of teaching novel skills to participants who are new to EAA. This study was also limited in that it only assessed the generalization of the skills the participants learned with the first author. Future research should seek to examine generalization of skills to peers as well as other familiar or unfamiliar adults.

There is a need for practitioners in the field of animal assisted activities and therapies to identify and effectively implement evidence-based practices that are most beneficial to their clients (Fine, 2006). The use of direct instruction and verbal prompting has been shown to be effective for teaching social skills (Rao, Beidel, & Murray, 2008) and there is potential for similar strategies to be used when working with animals in a variety of school and community settings. It may be of interest to examine the impact of other equine-assisted activities in school classrooms on teaching social skills through activities such as miniature horse programs. Future research could also examine how motivation and anxiety impact the teaching of social skills to children with disabilities and the effect of including animals.

References

Alberto, P. A., & Troutman, A. C. (2006). Applied behavior analysis for teachers. Upper Saddle River, NJ: Merrill/Prentice Hall.

American Psychiatric Association. (2013). Diagnostic and statistical manual of mental disorders (5th ed.). Washington, DC: Author

Anderson, S., & Meints, K. (2016). Brief report: The effects of equine-assisted activities on the social functioning in children and adolescents with autism spectrum disorder. *Journal of Autism and Developmental Disorders*, 46 (10), 3344-3352.

Barry, T. D., Klinger, L. G., Lee, J. M., Palardy, N., Gilmore, T., & Bodin, S.D. (2003). Examining the effectiveness of an outpatient clinic-based social skills group for high-functioning children with autism. *Journal of Autism and Developmental Disorders*, 33, 685–701.

Bass, M. M., Duchowny, C. A., & Llabre, M. M. (2009). The effect of therapeutic horseback riding on social functioning in children with autism. *Journal of Autism and Developmental Disorders*, 39, 1261-1267. doi: 10.1007/s10803-009-0734-3.

Bassette, L. A., Beck, A., Russell, M., Templin, S., & Brady, C. M. (2014). The effects of an equine-based mathematics curriculum on academic engagement behavior. *Scientific and Educational Journal of Therapeutic Riding*, 19, 26-41.

Bellini, S. (2008). Building social relationships: A systematic approach to teaching social interaction skills to children and adolescents with autism spectrum disorders. Paper presented to the Southern Indiana Career and Technical Center, Evansville, IN.

Clark, T., Feehan, C., Tinline, C., & Vostanis, P. (1999). Autistic symptoms in children with attention deficit-hyperactivity disorder. *European Child & Adolescent Psychiatry*, 8, 50-55.

de Boo, G. M. & Prins, P. J. M. (2006). Social incompetence in children with ADHD: Possible moderators and mediators in social-skills training. *Clinical Psychology Review*, 27, 78-97. doi:10.1016/j.cpr.2006.03.006.

Fine, A. H. (2006). Handbook on animal assisted therapy: Theoretical foundations and guidelines for practice. New York, NY: Academic Press.

Friedmann, E., Katcher, A.H., Thomas, S. A., Lynch, J. J., Messent, P. R., & Phil, D (1983). Social interaction and blood pressure: Influence of animal companions. *Journal of Nervous and Mental Disease*, 171, 461-465.

Gresham, F. M., Sugai, G., & Horner, R. H. (2001). Interpreting outcomes of social skills training for students with high incidence disabilities. *Exceptional Children*, 67, 331-344.

Kennedy, C. H. (2005). Single-case designs for educational research. Boston, MA: Allyn and Bacon.

Koenig, K. (2012). Practical social skills for autism spectrum disorders: Designing child specific interventions. New York, NY: Bytheway Publishing Services.

Kowalski, R. M., & Fedina, C. (2011). Cyber bullying in ADHD and Asperger syndrome populations. *Research in Autism Spectrum Disorders*, 5, 1201–1208, doi: 10.1016/j.rasd.2011.01.007

Martin, F., & Farnum, J. (2002). Animal-assisted therapy for children with pervasive developmental disorders. *Western Journal of Nursing Research*, 26, 657-670.

McAffe, J. (2002). Navigating the social world: A curriculum for individuals with Asperger's syndrome, high functioning autism, and related disorders. Arlington, TX: Future Horizons Inc.

PATH Intl. (2015). 2015 PATH Fact Sheet. Retrieved December 1, 2016 from http://www.pathintl.org

Rao, P. A., Beidel, D. C., & Murray, M. J. (2008). Social skills interventions for children with Asperger's syndrome or high-functioning autism: A review and recommendations. *Journal of Autism and Developmental Disorders*, 38, 353-361. doi: 10.1007/s10803-007-0402-4.

Rud, A., & Beck A. M. (2000). Kids and critters in class together. *Phi Delta Kappan*, 82, 313-315.

Sams, M. J., Fortney, E. V., & Willenbring, S. (2006). Occupational therapy incorporating animals for children with autism: A pilot investigation. *The American Journal of Occupational Therapy*, 60, 268-274.

Sansosti, F. J., & Powell-Smith, K. A. (2006a). Using social stories to improve the social behavior of children with Asperger syndrome. *Journal of Positive Behavior Interventions*, 8, 43–57.

Shiloh, S., Sorek, G., & Terkel, J. (2003). Reduction of state-anxiety by petting animals in a controlled laboratory experiment. *Anxiety, Stress & Coping*, 16, 387-395(9)

Smith, D. D., & Tyler, N. C. (2010). Introduction to special education: Making a difference (7th ed,). Columbus, OH: Pearson/Merrill.

Sterling, L., Dawson, G., Estes, A., & Greenson, J. (2008). Characteristics associated with presence of depressive symptoms in adults with autism spectrum disorder. *Journal of Autism and Developmental Disorders*, 38, 1011-1018. doi: 10.1007/s10803-007-0477-y.

Tse, J., Strulovitch, J., Tagalakis, V., Meng, L., & Fombonne, E. (2007). Social skills training for adolescents with Asperger syndrome and high-functioning autism. *Journal of Autism and Developmental Disorders*, 37, 1960-1968. doi 10.1007/s10803-006-0343-3.

ARTICLE THREE

The Unique Meaning of Equine Assisted Activities (EAA) for Riders, Parents, and Volunteers in a Therapeutic Riding Program

Kathryn M. Loukas, OTD, MS, OTR/L, FAOTA
Kirstin Russell, MS, OTR/L
Nicole Trifone, MS, OTR/L
Angela Brunette, MS, OTR/L
Josefine Balasz, MS, OTR/L
Laura Crossley-Marra, MS, OTR/L
Sarah Durgin, MS, OTR/L
Sarah Bronson, Executive Director

University of New England, Westbrook College of Health Professions,
Department of Occupational Therapy, and Riding to the Top Therapeutic Riding Center

Correspondence to:
Kathryn M. Loukas
Department of Occupational Therapy, University of New England
716 Stevens Ave.
Portland, ME 04103-2670
Email: kloukas@une.edu

THE UNIQUE MEANING OF EQUINE ASSISTED ACTIVITIES (EAA) FOR RIDERS, PARENTS, AND VOLUNTEERS IN A THERAPEUTIC RIDING PROGRAM

ABSTRACT

This qualitative study sought to identify the meaning individuals associate with equine-assisted activities (EAA), including riders, parents of riders, and volunteers participating in a therapeutic riding program through semi-structured interviews. Data was analyzed using phenomenological methods to find broad and role specific themes associated with EAA. Results suggest that doing, being, and belonging (Wilcock, 1998), all aspects of what the American Occupational Therapy Association defines as the daily life activities in which people engage, referred to as occupations (American Occupational Therapy Association, 2014), were common themes experienced among those participating in EAA.

KEYWORDS

Equine Assisted Activity; Therapeutic Riding; Meaningful Engagement; Purposeful Activity.

INTRODUCTION

Horses have shown the ability to provide individuals with unique and therapeutic experiences. According to Memishevikj and Hodzhikj (2010), the comprehensive goal of equine-assisted activities is to support a client's physical, emotional, cognitive, and social growth. EAA has proven to increase physical and psychosocial skills in participants with autism spectrum disorder, cerebral palsy, spinal cord injury, attention deficit and hyperactivity disorder, at-risk youth, and conduct disorder (Memishevikj & Hodzhiki, 2010; Kern et al., 2011; Bass, Duchowny, & Llabre, 2009; Debuse, Gibb & Chandler, 2009; Bowers, Ewing, MacDonald, & Taylor, 2007). In addition, it has facilitated increased attention and social skills in children with various learning and communicative disabilities (Cuypers, De Ridder, & Strandheim, 2011; Ghorban, Sedigheh, Marzieh, & Yaghoob, 2013).

When participating in EAA, the individual typically engages in activities that range from horseback riding to horsemanship skills on the ground. Horseback riding and horsemanship have the potential to be incorporated as a therapeutic medium to teach life skills to individuals with additional needs (Lee Young, 2005). In one study, the act of riding appeared to be a rewarding stimulus that accounted for higher levels of motivation, social engagement and participation in occupations as well as providing the riders with a schedule that helped them adapt and accept more flexible routines and lifestyles (Bass, Duchowny, & Llabre, 2009).

Masini (2010) described the process in which participants learn about themselves through their reaction to another living creature. Because of the size and noted sensitivity to human beings, horses make the perfect animal for this social transaction. Horses can serve as a medium where individuals can safely project uncomfortable feelings. During EAA there are opportunities to process interpersonal concerns, such as self-concept, respect, boundaries, assertiveness and trust.

Bowers, Ewing, MacDonald, and Taylor (2007) studied the use of horse programming with at-risk children. The authors found that horseback riding and horsemanship instills a sense of order, creates an understanding of boundaries, improves focus, and instills trust. Children with moderate to severe behavioral or conduct disorders and/or learning disabilities can benefit from horse academics, horse care and handling of the participant's partnered horse, and the opportunity to ride. Positive changes in conduct, social acceptance, trust, and confidence have been found following experiences with horses (All & Loving, 1999). Macauley and Gutierrez (2004) and Debuse, Gibb, and Chandler (2009) found equine assisted activities effective in improving self-concept, locus of control, affect, and behavior.

Rehabilitation professionals have previously examined the effectiveness of EAA, though there remains a lack of literature surrounding the meaning of EAA. The purpose of this study was to identify the common themes experienced by individuals in different roles associated with EAA.

Methods

This qualitative phenomenological research project was reviewed and approved by the University of New England's IRB committee. The team conducting the study consisted of eight investigators. Consent forms were approved and developed for each participant group; riders, parents, and volunteers. Those under the age of 18 were given consent to participate by a legal guardian. Semi structured interviews were completed with participants by pairs of researchers. Questions were developed for each subgroup role of each participant, as a parent, volunteer or rider. Open-ended questions allowed individuals to provide a qualitative narrative and interviewers facilitated the storytelling experience. All data was later analyzed to determine similar themes among the participants. Group comparisons and commonalities were then derived and analyzed.

Participants

Participants were recruited from a therapeutic riding center in New England through word of mouth, electronic and paper fliers, and response to requests. Four riders, six parents of riders, and five volunteers (n =15) participated in the study. To participate in the study, individuals had to be either riders at the riding center, parents of riders, or volunteers. Participants ranged in age from adolescent to older adults and included both males and females.

Setting

All subjects were interviewed at the therapeutic riding center. Convenient times were scheduled around times for lessons and volunteering.

Measures

This was a qualitative study with interview questions developed specifically for each group of participants (see appendix for interview questions).

PROCEDURE

All interviews were scheduled through the primary investigator and with the assistance of the director of the therapeutic riding center. Numbers were assigned to each participant in the order of dates of the interview, only the primary investigator had the original names of participants. Interviews lasted about 30 minutes and were performed with two members of the study team. One member of the pair conducted the interview while the other kept notes which were later transcribed.

DATA ANALYSIS

Transcribed interviews were de-identified and placed into electronic and paper theme books. Teams of three initially reviewed the transcription and began the thematic process. The study team met one afternoon with colored markers and large post-it papers to identify themes emerging from the data. Specific overarching quotes were used to support thematic findings; with supporting quotes listed below. All members of the study team participated in the thematic analysis and supported the following results.

RESULTS

THEMES FROM RIDERS:

"The total lack of judgment and the relationship with the horse is really powerful."

SAFETY

- Sense of protection on and around the horses
- "I like riding because I feel safe on the horse."
- "The horse is trusting."

SELF-ESTEEM

- "The horse is my idol and teacher, she made me do better."
- "It emphasizes my strengths more than weaknesses."

CONNECTION

- "There is something about relating to the horse that really helped me feel whole again."

LOVE OF THE HORSES AND THE EXPERIENCE

- "Horses do an amazing job of allowing me to have a good experience."
- "Be yourself and they will take care of you."

COURAGE

- "The horses give me courage to speak up and do better."

THEMES FROM PARENTS:

"My child gains more confidence here than anywhere else."

RELATIONSHIPS

- "It's just a different type of connection."

SELF-WORTH

- "My son feels so empowered."

CONFIDENCE

- "When [my child] is riding, she feels big and strong."
- "This is where my child found her physical confidence."

PRIDE

- "I respect [my child's] ability."

MINDFULNESS

- "He is very focused, pays attention, engaged, and always happy to ride."

IDENTITY BUILDING

- "She's a horse girl."

Multisensory and natural

- "The smells and movements are so much deeper."
- "The range of stimuli that she is receiving. It is amazing."

Described in one word:

- "Empowerment"
- "Respect"
- "Sense of Importance"

Themes from volunteers

"Little miracle moments we get to see all the time."

Altruistic and non-altruistic

- Giving and receiving
- "It gives me a lot. It really is a two way street, not just altruistic."
- "I ultimately grew as a person by being involved."

Collaboration

- It's a multi-faceted experience
- "You really get something from the animal, instructor and the rider."

Transformative experiences

- "The interaction, the kindness, the care, and the compassion speaks to my soul."
- "Watching the rider get connected to the horse and seeing them improve is something I can always take with me."

Progress

- "This program is totally dedicated to meeting goals. They have a strategic plan and are very involved. They revisit their goals and constantly revise and improve. Everyone is willing to do the smallest task, which is what has made this place so successful."

IDENTITY

- "Giving a little of myself back to the community."
- "The inner peace that I can take with me."
- "Out of all my years out here I have never seen anyone not affirmed. The horses provide such motivation."

OVERALL THEMES

Self-worth: Relationships: Identity Building (Table 1).

Table 1: Participant Information

Overall Themes:	***Self-worth; Relationships; Identity Building***
Riders	Safety; Self-esteem; Connection; Courage
Parents	Relationships; Self-worth; Confidence; Pride; Mindfulness; Identity building; Multisensory and natural
Volunteers	Little Miracles; Collaboration; Transformative experiences; Progress; Identity

DISCUSSION AND CONCLUSION

This small, qualitative study supports the idea that EAA provides meaningful experiences to groups beyond the rider. In the results of this study, the transformative themes of meaningful occupations; *doing, being, belonging* and *becoming* were found (see also American Occupational Therapy Association, 2014, for a definition of the term "occupation" within occupational therapy). Doing refers to the sense of purpose, fulfillment, and ability to engage in personally meaningful experiences (Hammell, 2004). Experience is more than just doing, it is a "synthesis of doing, being and becoming" (Wilcock, 1998a, p. 341). By engaging in EAA, individuals can find a sense of purpose, fulfillment, and an affirmation of competence with an enhanced feeling of self-worth. Through the process of doing, an individual can develop a sense of being by reflecting, (re)discovering the self, savoring and appreciating the moment (Hammell, 2004). While expressing meaning through being, an individual can find a sense of belonging through contributing socially to others involved in EAA. Social interaction, mutual support, and friendship help foster a sense of affirmation and connectedness for individuals (Robeiro et al., 2001). With increased security and sense of belonging, an individual can envision a journey of becoming which enhances the vision of a future self and possible lives (Hammell, 1998a, 2003a). EAA allows individuals the opportunity to embark on a journey through doing activities, being a horseback rider, belonging to a community, and becoming whatever they dream to be.

Limitations of the study include the small sample size and a single facility. The findings of this study support the premise of doing, being, becoming and belonging in EAA as meaningful engagement in the experiences of riders, parents, and volunteers at one therapeutic riding facility.

References

All, A. C., & Loving, G. L. (1999). Animals, horseback riding, and implications for rehabilitation therapy. Journal of Rehabilitation, 65(3), 49-57.

American Occupational Therapy Association. (2014). Occupational therapy practice framework: Domain and process (3rd ed.). American Journal of Occupational Therapy, 68(Suppl. 1), S1-S48.

Bass, M. M., Duchowny, C. A., & Llabre, M. M. (2009). The effect of therapeutic horseback riding on social functioning in children with autism. The Journal of Autism and Developmental Disorders, 39(9), 1261-1267.

Bowers, M.J., Ewing, C.A., MacDonald, P.M., & Taylor, M. (2007). Equine-facilitated learning for youths with severe emotional disorders: A quantitative and qualitative study. Child Youth Care Forum, 36, 59-72.

Debuse, D., Gibb, C., & Chandler, C. (2009). Effects of hippotherapy on people with cerebral palsy from the users' perspective: A qualitative study. Physiotherapy Theory and Practice, 25(3), 174-192.

Cuypers, K., De Ridder, K., & Strandheim, A. (2011). The effect of therapeutic horseback riding on 5 children with attention deficit hyperactivity disorder: a pilot study. J Altern Complement Med, 17(10):901-8. doi: 10.1089/acm.2010.0547.

Ghorban, H., Sedigheh, R. D., Marzieh, G., & Yaghoob, G. (2013). Effectiveness of therapeutic horseback riding on social skills of children with autism spectrum disorder in Shiraz, Iran. Journal of Education and Learning, 2(3), 79-84.

Hammell, K. W. (2004). Dimmensions of meaning in the occupations of daily life. Canadian Journal of Occupational Therapy, 71(5), 296-305.

Kern, J. K., Fletcher, C. L., Garver, C. R., Mehta, J. A., Grannemann, B. D., Knox, K. R., & Trivedi, M. H. (2011). Prospective trial of equine-assisted activities in autism spectrum disorder. Alternative Therapies in Health and Medicine, 17(3), 14-20.

Lee Young, R. (2005). Horsemastership part 1 : Therapeutic components and link to occupational therapy. Internal Journal of Therapy and Rehabilitation, 12(2), 78-82.

Macauley, B. L., & Gutierrez, K. M. (2004). The effectiveness of hippotherapy for children with language-learning disabilities. Communication Disorders Quarterly, 25(4), 205-217.

Masini, A. (2010). Equine-assisted psychotherapy in clinical practice. Journal of Psychosocial Nursing And Mental Health Services, 48(10), 30-34.

Memishevikj, H., & Hodzhikj, S. (2010). The effects of equine assisted therapy in improving the psychosocial functioning of children with autism. The Journal of Special Education and Rehabilitation, 11(3), 57-67.

Robeiro, K. L., Day, D., Semeniuk, B., O'Brien, M., & Wilson, B. (2001). Northern initiative for social action: An occupation-based mental health program. American Journal of Occupational Therapy, 55, 493-500.

Wilcock, A. A. (1998). Occupation for health. British Journal of Occupational Therapy, 61, 340-345.

Appendix: Interview Questions

Interview questions for the Rider:

1. What is (are) the horses' name you ride? Tell us about your horse.
2. Tell me three interesting things about yourself.
3. What do you like to do for fun?
4. How long have you been riding horses?
5. What do you like about coming to Riding to the Top (RTT)?
6. Tell me what it is like to ride a horse? How does it make you feel?
7. What are the best parts of RTT? (horses? helpers?)
8. Have you made any friends at RTT?
9. Tell me about the helpers at RTT.
10. What would you tell my friend who is coming to RTT for the first time?

Interview questions for the Parent:

1. How long have you and your family member been coming to RTT?
2. What attracted you to bring your child to RTT?
3. Are there any changes you observe in your family member from before the session to after?
4. Are there changes in the family that you attribute to participating in the therapeutic horseback riding?
5. What would you say is something the family as the whole has gained from RTT?
6. Does your family member express a sense of importance from being able to participate in this program, in any way?
7. Do you see a change in your family member when on the horse as compared to doing activities off of them?
8. Do you feel that your family member has gained a better sense of self since starting these lessons?
9. In your opinion, how does therapeutic horseback riding relate to traditional therapy in terms of personal meaning?
10. In one word or phrase, how would you describe the experience of RTT to your family member or yourself?

INTERVIEW QUESTIONS FOR VOLUNTEERS:

1. Have you had any previous volunteer experience?
2. How long have you been a volunteer at RTT?
3. Tell us about your responsibilities associated with being a volunteer at RTT.
4. Have you had previous experience with horses and if so what?
5. What do you enjoy most about volunteering in equine assisted experiences?
6. What is unique about the RTT volunteer experience?
7. Would you recommend RTT to friends looking for volunteer opportunities?
8. In one word or phrase what does volunteering here mean to you?
9. Is there anything you would change about the program?
10. Why did you originally choose to volunteer at RTT?
11. Are you currently working?
12. How many hours a week do you volunteer here?

THE UNIQUE MEANING OF EQUINE ASSISTED ACTIVITIES (EAA) FOR RIDERS, PARENTS, AND VOLUNTEERS IN A THERAPEUTIC RIDING PROGRAM

ARTICLE FOUR

The Effect of Hippotherapy on Static and Dynamic Balance and Reaction Time in Autism Spectrum Disorder

Sara Mirzabeigi Fini, MS, Young Researchers and Elite Club, Islamshahr Branch, Islamic Azad University
Yaser Kazemzadeh, PhD, Department of Physical Education and Sport Sciences, Faculty of Physical Education, Islamic Azad University
Yahya Sokhanguei. PhD, Department of Physiotherapy, University of Social Welfare and Rehabilitation Sciences

Correspondence to:
Sara Mirzabeigi Fini, Young Researchers and Elite Club,
Islamshahr Branch, Islamic Azad University,
Islamshahr, Iran, No.3, South Motahari, Darya Blvd. Saadat Abaad Street,
1466944134, Tehran, Iran
E-mail: sarafini@iiau.ac.ir

Sources of support:
This study was supported by grants from the research Foundation of the Committee of Veterans and Disabled Equestrian Federation of Iran and the Department of Physical Education of Islamic Azad University Islamshahr branch.

THE EFFECT OF HIPPOTHERAPY ON STATIC AND DYNAMIC BALANCE AND REACTION TIME IN AUTISM SPECTRUM DISORDER

ABSTRACT

The term hippotherapy refers to how occupational therapy, physical therapy, and speech-language pathology professionals use evidence-based practice and clinical reasoning in the purposeful manipulation of equine movement to engage sensory, neuromotor, and cognitive systems to achieve functional outcomes (American Hippotherapy Association, 2016). This paper discusses the effect of 14 weeks of hippotherapy on static balance, dynamic balance and reaction time of individuals with autism spectrum disorder. The participants consisted of 24 males between 7-18 years old, diagnosed with autism spectrum disorder and randomly placed in experimental and control groups (N=12 for each group). The experimental group participated in 14 weeks of hippotherapy sessions. The control group did not engage in any hippotherapy or any physical activity outside of their ordinary life during the research period. The statistical analysis shows significance in the scores of the static and dynamic balance of the experimental group versus the control group as compared with the time before the exercises (p=0.0001), (p=0.019). In contrast, no significant change was seen in the reaction time scores of the participants as compared with the time before hippotherapy intervention (p=0.123). The result shows that 14 sessions of hippotherapy had a positive effect on static and dynamic balance of people diagnosed with autism spectrum disorder, with improvement in balance systems and physical statue, especially flexibility and strength, but shows no specific changes in the score of reaction time.

KEYWORDS

Autism; Dynamic Balance; Hippotherapy; Reaction Time; Static Balance.

INTRODUCTION

Autism is a disorder of social relations, which is characterized by inability in verbal and nonverbal interpersonal relations and having imaginative activities and limited interests and activities (Fombonne & Tidmarsh, 2003). Repeated movements (clapping, jumping), unusual responses to people, attachment to objects and/or resistance to changes are seen in individuals with autism spectrum disorder (Pan & Frey, 2006). Autism is a permanent and lifelong disorder and the rate of autism in males is 3-4 times higher than the one for females (Blenner & Augustyn, 2014; Dodd, 2005). However, there is evidence that some methods, including behavioral treatment such as applied behavior analysis (ABA) or consumption of medicines such as Risperidone and Aripiprazole, can lead to reduction of behavioral symptoms such as irritability and/or aggression (Baron-Cohen et al., 2009; Tonge, Bull, Brereton, & Wilson, 2014; Boon-yasidhi, Jearnarongrit, Tulayapichitchock, & Tarugsa, 2014; Ghaleiha et al., 2015). Uncertain results of behavioral treatments and severe side effects of drugs motivate families of children with Autism to look for complementary and alternative medicine treatments (Ishitobi et al., 2013).

Hippotherapy is one of the most well known categories of animal-assisted therapy. (Stanley-Hermanns & Miller, 2002). According to The American Hippotherapy Association, the term hippotherapy refers to how occupational therapy, physical therapy, and speech-language pathology professionals use evidence-based practice and clinical reasoning in the purposeful manipulation of equine movement to engage sensory, neuromotor, and cognitive systems to achieve functional outcomes (American Hippotherapy Association, 2016). In fact, horses impose unique neuromuscular simulation on humans when riding. Rhythmic movements of a horse transmitted to a rider are similar to the movement of a human when walking. (Thompson et al., 2014). The ability to control speed of a horse allows adjustment of the rider's pelvic movement. Hippotherapy and the use of the movement of the horse can affect balance, coordination, flexibility and self-confidence in a rider (Borzo, 2002; Borgi et al., 2016; Holm et al., 2014; Silkwood-Sherer, Killian, Long, & Martin, 2012; Bronson, Brewerton, Ong, Palanca, & Sullivan, 2010).

Various training games and physical activities can be used to decrease stereotypical behavior and increase vital social skills of children with autism. In fact, balance control can provide proper conditions for daily performance improvement (Maurice, Mannion, Letso, & Perry, 2001). The balance control system requires three kinds of sensory input: visual, vestibular and proprioceptive. Generally, the dynamic sense of balance depends primarily on vestibular input on unstable surfaces and static balance depends primarily on somatosensory input (Borzo, 2002).

THE EFFECT OF HIPPOTHERAPY ON STATIC AND DYNAMIC BALANCE AND REACTION TIME IN AUTISM SPECTRUM DISORDER

Hippotherapy is an intervention that allows therapists to work on improving activity performance, while the child participates in what they might consider a fun and recreational activity. As an intervention, hippotherapy can be highly motivating and provides an opportunity for children to challenge themselves and work harder than they might ordinarily do in more traditional therapy settings (Borges, Werneck, Silva, Gandolfi, & Pratesi, 2011). In hippotherapy, therapists use horse movements as a therapeutic tool for improving functional limitations in patients (Lee, Kim, & Na, 2014ab).

Exercises used in hippotherapy may help improve balance, coordination, and posture through strengthening and developing motor, sensory, and perceptual skills (Thompson, Ketcham, & Hall, 2014; Champagne & Dugas, 2010). It seems that the results stem from the formation of new and involuntary contractions in muscles because of exposure to new and moving surfaces and that the contractions help maintain trunk position and balance at that moment (Sterba, Rogers, France, & Vokes, 2002; MacPhail et al., 1998). In addition, unbalanced surfaces can improve an individual's ability to stabilize (Sterba et al., 2002; MacPhail et al., 1998). The rhythmic movements of a horse may facilitate quality of standing and balance of patients (McGibbon, Andrade, Widener, & Cintas, 1998; Sterba et al., 2002; Cherng, Liao, Leung, & Hwang, 2004; Shurtleff, Standeven, & Engsberg, 2009). The body's balance system is strengthened, which leads to improved balance and coordination (Sterba et al., 2002). Doing exercises on the back of a horse can strengthen riders' pelvic, hip, trunk and spine muscles (Hadders-Algra, Brogren, & Forssberg, 1998). Snider, Korner-Bitensky, Kammann, Warner and Saleh (2007) studied children with cerebral palsy and announced that hippotherapy had short-term effects on the improvement of symmetry of trunk and pelvic muscles, whereas hippotherapy exercises had no effect on muscle tone improvement.

The most important benefit of hippotherapy is that the trainer has the ability to regulate the walking speed of the horse, which helps facilitate the pelvic movement of the rider. Several studies show that hippotherapy might be considered a useful method for improving flexibility (Benda, McGibbon, & Grant, 2003; Shurtleff et al., 2009; Lechner, Kakebeeke, Hegemann, & Baumberger, 2007; Giagazoglou, Arabatzi, Dipla, Liga, & Kellis, 2012) and muscle strength (Benda et al., 2003; Manikowska, Jóźwiak, Idzior, Chen, & Tarnowski, 2013) of patients. In fact, improvement of the physical condition of patients may enable their bodies to maintain a stable and balanced condition (Manikowska et al., 2013). Through measuring the relationship between gross motor skills and social behavior of the children with autism spectrum disorder, Debuse, Gibb, and Chandler (2009) concluded that the better the motor skills of children with autism are, the higher their social competence will be.

This study is aimed at discussing the effect of hippotherapy on static and dynamic balance and reaction time of individuals with autism spectrum disorder.

METHOD

Participants in this study were 24 individuals between the ages of 7 and 18 living in Tehran, Iran, who were diagnosed with autism spectrum disorder and under treatment by psychotherapists and occupational therapists in private accredited centers. They were selected based on convenience. Participants were randomly placed in experimental and control groups (N=12 for each group). The experimental group participated in 14 weeks of weekly hippotherapy sessions. The control group did not perform any specific physical activities during the time of the study. The inclusion criteria were: no record of epilepsy, orthopedic, cardiovascular, or respiratory diseases. Prior to the study all candidates were asked to provide completed personal and medical forms in which they show their consent to participate in the study.

Rates for static and dynamic balance and reaction time of both groups were measured before and after the intervention. Attempts were made to select simple and understandable tests for the population of this study. The Stork Balance Stand test (Johnson & Nelson, 1979) was used for measuring static balance. In the test the participant stands barefoot on a flat, immovable surface with eyes open. The participant places their hands on their hips and when cued, lifts their dominant foot and places it beside the non-dominant knee. A digital chronometer is used for calculating the time the participant keeps their dominant foot elevated. The time the participant is able to maintain this position is recorded as their score.

A balance board was used for evaluating and measuring dynamic balance. In this part of the test, the participant is asked to stand barefoot on the balance board. Upon a prompt, the participant attempts to establish balance without touching the ground with the edges of the balance board or any part of their body. When this occurs the test is stopped and the time is recorded for the individual.

For measuring reaction time, the Nelson's Speed of Movement test was employed (Mohammadpour, Hashemian, Yaghoubi, & Valiolah, 2012), using a 50 cm ruler covered by a color tape. The participant is requested to sit on a specific place and hold their hands facing each other in front of their chest with a distance of 30 centimeters between them. The ruler is droppedand the participant is asked to prevent it from falling on the floor by clasping it between their hands. Letting the ruler fall on the floor and the use of other body parts to hold it constitute an error. The measure on the ruler touching the hands of the participant is entered into their record.

To familiarize participants with each test they were allowed to perform three pilot tests. The next three times were measured and recorded as the records of the individuals from both, the control and the experimental groups. Between each performance there was a five-second interval.

The intervention protocol includes some specific exercises, which were done beside a horse and while horseback riding. The exercises were done during 60 to 90 minute sessions for 14 successive weeks, which were intensified over time. It should be noted that a pediatric physical therapist and an occupational therapist accompanied the team during the study. Table 1 shows the intervention protocol for the 14 sessions of hippotherapy exercises.

Table 1. The lesson plan of 14 sessions of Hippotherapy exercises

The Lesson Plan of Hippotherapy Program: 14 Sessions of exercises	
First Phase - 4 Sessions	
Communicating with a horse (patting, currycombing, and feeding)	15 min
Stretching and flexibility movements on a horse	10 min
Major exercises while horseback riding: **1.** Balance exercises (hands foreword, on hips, at sides) **2.** Exercises using a big ring (around waist, above head) **3.** Exercises of hitting trainer's hands (front, rear, back)	25min
Games and amusement, free movements in a riding arena, singing while horseback riding	10 min
Second Phase - 4 Sessions	
1. Communicating with a horse (patting, currycombing, and feeding) **2.** Stretching and flexibility movements on a horse	20 min
Major exercises during horseback riding: **1.** Balance exercises (hands foreword, on hips, at sides) **2.** Standing on stirrups and lying on a horse **3.** Exercises using small rings (around hands at sides, above head, front) **4.** Hitting target using a wooden sword randomly determined by trainer (front, rear, back) **5.** Exercises of holding a ball (in front of chest, above head, sides by two hands, rear)	45 min
Games and amusement, free movements in a riding arena, singing while horseback riding	10 min

Table 1. The lesson plan of 14 sessions of Hippotherapy exercises – cont'd.

The Lesson Plan of Hippotherapy Program: 14 Sessions of exercises	
Third Phase - 4 Sessions	
1. Communicating with a horse (patting, currycombing, and feeding) **2.** Stretching and flexibility movements on a horse	20 min
Major exercises while horseback riding: **1.** Standing on stirrups and lying on a horse **2.** Exercises using sports cones and throwing rings (front, back, above) **3.** Closing hula hoop rings **4.** Hitting a target using a wooden sword randomly determined by trainer (front, rear, back) **5.** Throwing a ball to trainer, vice versa	55 min
Games and amusement, free movements in a riding arena, singing while horseback riding	10 min
Fourth Phase - 2 Sessions	
1. Communicating with a horse (patting, currycombing, and feeding) **2.** Stretching and flexibility movements on a horse	15 min
Major exercises while horseback riding: **1.** Exercises using sports cones and throwing rings (front, back, above) **2.** Exercises of throwing and receiving a ball **3.** Exercises of closing hula hoop rings on special bases in a riding arena **4.** Exercises of throwing ball to a basketball ring **5.** Identifying and putting color beads into a special cube	65 min
Games and amusement, free movements in a riding arena, singing while horseback riding	10 min

RESULTS

Analysis of covariance was used for comparing the pre-test and post-test of the control and the experimental groups. Significance level was $\alpha \leq 0.05$. All the statistical calculations were made in SPSS Version 21.

Table 2 shows the mean and SD of pre-tests and post-tests for static balance, dynamic balance, and reaction time scores.

Table 2. Minimum, maximum, mean, and SD of age, height, and weight variables

Variables	Minimum	Maximum	Mean	SD
Age(year)	7	18	11.08	3/03
Height(cm)	122	168	146/2	12/4
Weight(kg)	24	81	45/5	13/9

Figures 1, 2, and 3 show the diagram of research variables based on the mean scores in pre-test and post-test for experimental and control groups. Results in Figure 1 show a significant improvement in mean scores of static balance between pre-test (7.88) and post-test (11.5). In addition, the dramatic progress can be seen on the mean scores of dynamic balance between pre-test (10.02) and post-test (12.8) in Figure 2. Finally, Figure 3 shows there is no improvement in the mean scores on reaction time before intervention (0.26) and after (0.23).

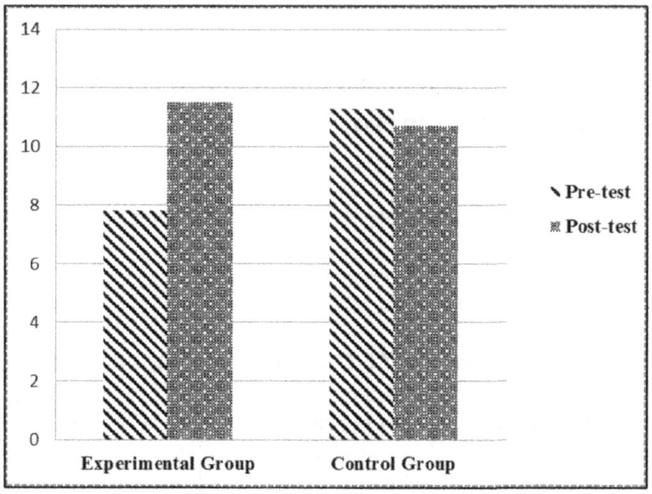

Figure 1. Diagram of the mean of static balance (in seconds) in both research groups in pre-test and post-test.

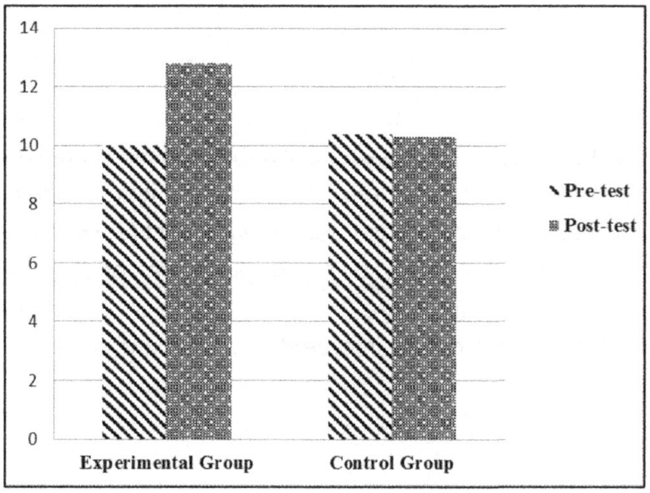

Figure 2. Diagram of mean of dynamic balance (in seconds) for both research groups in pre-test and post-test

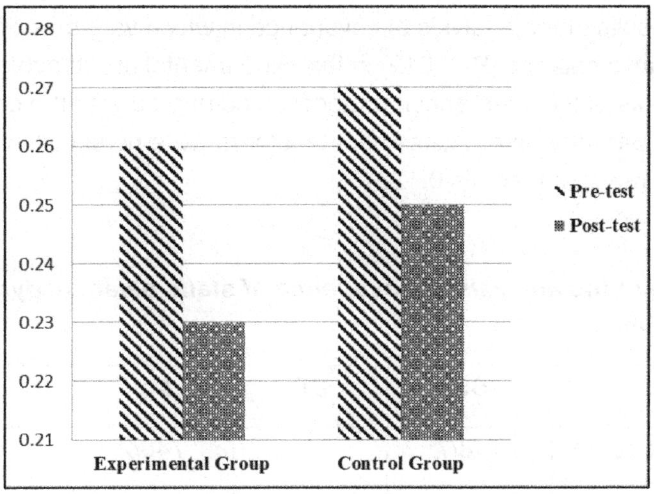

Figure 3. Diagram of mean of reaction time (in milliseconds) for both research groups in pre-test and post-test

Table 3 shows the mean and standard deviation of pre-test and post-test of static balance, dynamic balance, and reaction time scores.

Table 3. Description of research variables

Variables	Level	Hippotherapy		Control	
		Mean	SD	Mean	SD
Static balance	pre-test	7/88	6/2	11/3	5/2
	post-test	11/5	7/2	10/7	5/03
Dynamic balance	pre-test	10/02	4/6	10/4	3/5
	post-test	12/8	3/9	10/3	3/5
Reaction time	pre-test	0/26	0/06	0/27	0/02
	post-test	0/23	0/04	0/25	0/03

Table 4 shows the results of the analysis of covariance in which variables of static balance (P=0.0001) and dynamic balance (P=0.019) in the experimental group achieved a significant increase after 14 weeks of the intervention protocol as compared with the control group. In contrast, it became clear after data assessment that there were no significant differences in the score of participants' reaction time (P=0.123).

Table 4. The results of the analysis of covariance of static balance, dynamic balance, and reaction time variables

Source		SS	DF	MS	F	Sig.
Static balance	Pre-test	748/7	1	748/2	151/1	0/000
	Group	100/7	1	100/7	20/3	0/001
	Error	103/9	21	4/95		
Dynamic balance	Pre-test	158/9	1	158/9	22/07	0/000
	Group	46/5	1	46/5	6/47	0/019
	Error	151/1	21	7/2		
Reaction time	Pre-test	0/025	1	0/025	50/7	0/000
	Group	0/001	1	0/001	2/57	0/123
	Error	21	21	0/000		

DISCUSSION

This paper discussed the effect of 14 weeks of hippotherapy sessions on static balance, dynamic balance, and reaction time of individuals with autism spectrum disorder. The statistical analysis showed significant progress in the scores of the static and dynamic balance of the experimental group versus the control group as compared with the time before exercises. In contrast, the research results indicated that the horseback riding exercises had no effect on the participants' reaction time. The research results were inconsistent with the research of Giagazoglou et al. (2013).

Until recently, the use of the horse in therapy had not been seen as a beneficial option for children with developmental delays, motor deficits, and neuropsychiatric disorders such as autism (Marcs et al., 2012). However, the results of the present study provide evidence that hippotherapy should be incorporated into these children's therapeutic programs. Results were consistent with those of similar studies (Lee et al., 2014ab; Zadnikar & Kastrin, 2011).

ACKNOWLEDGEMENTS

The authors want to thank all the participants for their contributions as well as all the staff at the Physical Education departments of the Islamic Azad University, Islamshahr branch, and the Committee of Veterans and Disabled Equestrian Federation of Iran. Thanks to Dr. Siavash Salek Ardestani who assisted in this project and whose guidance was valuable.

REFERENCES

American Hippotherapy Association (2016). "What is Hippotherapy". Retrieved from: http://www.americanhippotherapyassociation.org

Baron-Cohen, S., Scott, F. J., Allison, C., Williams, J., Bolton, P., Matthews, F. E., & Brayne, C. (2009). Prevalence of autism-spectrum conditions: UK school-based population study. *The British Journal of Psychiatry*, 194 (6), 500-509.

Benda, W., McGibbon, N. H., & Grant, K. L. (2003). Improvements in muscle symmetry in children with cerebral palsy after equine-assisted therapy (hippotherapy). *The Journal of Alternative & Complementary Medicine*, 9 (6), 817-825.

Blenner, S., & Augustyn, M. (2014). Is the prevalence of autism increasing in the United States? *BMJ*, 348, g3088.

Boon-yasidhi, V., Jearnarongrit, P., Tulayapichitchock, P., & Tarugsa, J. (2014). Adverse effects of risperidone in children with autism spectrum disorders in a naturalistic clinical setting at Siriraj Hospital, Thailand. *Psychiatry Journal.* doi: 10.1155/2014/136158

Borges, M. B. S., Werneck, M. J. D. S., Silva, M. D. L. D., Gandolfi, L., & Pratesi, R. (2011). Therapeutic effects of a horse riding simulator in children with cerebral palsy. *Arquivos de Neuro-Psiquiatria*, 69 (5), 799-804.

Borgi, M., Loliva, D., Cerino, S., Chiarotti, F., Venerosi, A., Bramini, M., & Bisacco, F. (2016). Effectiveness of a standardized equine-assisted therapy program for children with autism spectrum disorder. *Journal of Autism and Developmental Disorders*, 46 (1), 1-9.

Borzo, G. (2002). Horsepower. *American Medical News*, 45 (23), 24-26.

Bronson, C., Brewerton, K., Ong, J., Palanca, C., & Sullivan, S. J. (2010). Does hippotherapy improve balance in persons with multiple sclerosis: a systematic review. *European Journal of Physical and Rehabilitation Medicine*, 46 (3), 347-353.

Cerquozzi, C., Cerquozzi, E., Darragh, A., & Miller-Kuhaneck, H. (2007). An exploratory survey of occupational therapists' role in hippotherapy. *Occupational Therapy Faculty Publications 18.*

Champagne, D., & Dugas, C. (2010). Improving gross motor function and postural control with hippotherapy in children with Down syndrome: Case reports. *Physiotherapy Theory and Practice*, 26 (8), 564-571.

Cherng, R. J., Liao, H. F., Leung, H. W., & Hwang, A. W. (2004). The effectiveness of therapeutic horseback riding in children with spastic cerebral palsy. *Adapted Physical Activity Quarterly*, 21(2), 103-121.

Debuse, D., Gibb, C., & Chandler, C. (2009). Effects of hippotherapy on people with cerebral palsy from the users' perspective: A qualitative study. *Physiotherapy Theory and Practice*, 25 (3), 174-192.

Dodd, S. (2005). Understanding autism. Marrickville, N.S.W.: Elsevier Australia.

Fombonne, E., & Tidmarsh, L. (2003). Epidemiologic data on Asperger disorder. *Child and Adolescent Psychiatric Clinics of North America*, 12 (1), 15-21.

Ghaleiha, A., Rasa, S. M., Nikoo, M., Farokhnia, M., Mohammadi, M. R., & Akhondzadeh, S. (2015). A pilot double-blind placebo-controlled trial of pioglitazone as adjunctive treatment to risperidone: Effects on aberrant behavior in children with autism. *Psychiatry Research*, 229(1), 181-187.

Giagazoglou, P., Arabatzi, F., Dipla, K., Liga, M., & Kellis, E. (2012). Effect of a hippotherapy intervention program on static balance and strength in adolescents with intellectual disabilities. *Research in Developmental Disabilities*, 33 (6), 2265-2270.

Giagazoglou, P., Arabatzi, F., Kellis, E., Liga, M., Karra, C., & Amiridis, I. (2013). Muscle reaction function of individuals with intellectual disabilities may be improved through therapeutic use of a horse. *Research in Developmental Disabilities*, 34 (9), 2442-2448.

Granados, A. C., & Agís, I. F. (2011). Why children with special needs feel better with hippotherapy sessions: a conceptual review. *The Journal of Alternative and Complementary Medicine*, 17 (3), 191-197.

Hadders-Algra, M., Brogren, E., & Forssberg, H. (1998). Development of postural control—differences between ventral and dorsal muscles? *Neuroscience & Biobehavioral Reviews*, 22 (4), 501-506.

Holm, M. B., Baird, J. M., Kim, Y. J., Rajora, K. B., D'Silva, D., Podolinsky, L., ... & Minshew, N. (2014). Therapeutic Horseback Riding Outcomes of Parent-Identified Goals for Children with Autism Spectrum Disorder: An ABA' Multiple Case Design Examining Dosing and Generalization to the Home and Community. *Journal of Autism and Developmental Disorders*, 44 (4), 937-947.

Ishitobi, M., Kosaka, H., Takahashi, T., Yatuga, C., Asano, M., Tanaka, Y, ... & Tomoda, A. (2013). Effectiveness and tolerability of switching to aripiprazole from risperidone in subjects with autism spectrum disorders: a prospective open-label study. *Clinical Neuropharmacology*, 36 (5), 151-156.

Johnson, B., & Nelson, J. (1979). Practical measurements for evaluation in physical education (4th edition). Minneapolis, MN: Burgess

Lechner, H. E., Kakebeeke, T. H., Hegemann, D., & Baumberger, M. (2007). The effect of hippotherapy on spasticity and on mental well-being of persons with spinal cord injury. *Archives of Physical Medicine and Rehabilitation*, 88 (10), 1241-1248.

Lee, C. W., Kim, S. G., & Na, S. S. (2014a). The effects of hippotherapy and a horse riding simulator on the balance of children with cerebral palsy. *Journal of Physical Therapy Science*, 26 (3), 423-425.

Lee, C. W., Kim, S. G., & Yong, M. S. (2014b). Effects of hippotherapy on recovery of gait and balance ability in patients with stroke. *Journal of Physical Therapy Science*, 26 (2), 309-311.

MacPhail, H. A., Edwards, J., Golding, J., Miller, K., Mosier, C., & Zwiers, T. (1998). Trunk postural reactions in children with and without cerebral palsy during therapeutic horseback riding. *Pediatric Physical Therapy*, 10 (4), 143-147.

Mohammadpour, M., Hashemian, S., Yaghoubi, H., & Valiolah, S. (2012). The effect of mental practice on response time via Nelson's speed of movement test. *European Journal of Experimental Biology*, 2 (6), 2444-2450

Manikowska, F., Jóźwiak, M., Idzior, M., Chen, P. J., & Tarnowski, D. (2013). The effect of a hippotherapy session on spatiotemporal parameters of gait in children with cerebral palsy-pilot study. *Ortopedia, Traumatologia, Rehabilitacja*, 15 (3), 253-257.

Marcus, D. A., Bernstein, C. D., Constantin, J. M., Kunkel, F. A., Breuer, P., & Hanlon, R. B. (2012). Animal-assisted therapy at an outpatient pain management clinic. *Pain Medicine*, 13 (1), 45-57.

Maurice, C., Mannion, K., Letso, S., & Perry, L. (2001). Parent voices: Difficulty in accessing behavioral intervention for autism; working toward solutions. *Behavioral Interventions*, 16 (3), 147-165.

McGibbon, N. H., Andrade, C. K., Widener, G., & Cintas, H. L. (1998). Effect of an equine-movement therapy program on gait, energy expenditure, and motor function in children with spastic cerebral palsy: A pilot study. *Developmental Medicine & Child Neurology*, 40 (11), 754-762.

Nordin, V., & Gillberg, C. (1996). Autism spectrum disorders in children with physical or mental disability or both. I: Clinical and epidemiological aspects. *Developmental Medicine & Child Neurology*, 38 (4), 297-313.

Pan, C. Y., & Frey, G. C. (2006). Physical activity patterns in youth with autism spectrum disorders. *Journal of Autism and Developmental Disorders*, 36 (5), 597-606.

Peterka, R. J., & Loughlin, P. J. (2004). Dynamic regulation of sensorimotor integration in human postural control. *Journal of Neurophysiology*, 91 (1), 410-423.

Shumway-Cook, A., & Woollacott, M. H. (2007). Motor control: translating research into clinical practice. Philadelphia, PA: Lippincott Williams & Wilkins.

Shurtleff, T. L., Standeven, J. W., & Engsberg, J. R. (2009). Changes in dynamic trunk/head stability and functional reach after hippotherapy. *Archives of Physical Medicine and Rehabilitation*, 90 (7), 1185-1195.

Silkwood-Sherer, D. J., Killian, C. B., Long, T. M., & Martin, K. S. (2012). Hippotherapy—an intervention to habilitate balance deficits in children with movement disorders: a clinical trial. *Physical Therapy*, 92 (5), 707-717.

Snider, L., Korner-Bitensky, N., Kammann, C., Warner, S., & Saleh, M. (2007). Horseback riding as therapy for children with cerebral palsy: is there evidence of its effectiveness? *Physical & Occupational Therapy in Pediatrics*, 27 (2), 5-23.

Stanley-Hermanns, M., & Miller, J. (2002). Animal-Assisted Therapy: Domestic animals aren't merely pets. To some, they can be healers. *AJN, American Journal of Nursing*, 102 (10), 69-76.

Sterba, J. A., Rogers, B. T., France, A. P., & Vokes, D. A. (2002). Horseback riding in children with cerebral palsy: effect on gross motor function. *Developmental Medicine & Child Neurology*, 44 (05), 301-308.

Thompson, F., Ketcham, C. J., & Hall, E. E. (2014). Hippotherapy in children with developmental delays: Physical function and psychological benefits. *Advances in Physical Education*, 4, 60-69.

Tonge, B. J., Bull, K., Brereton, A., & Wilson, R. (2014). A review of evidence-based early intervention for behavioural problems in children with autism spectrum disorder: the core components of effective programs, child-focused interventions and comprehensive treatment models. *Current Opinion in Psychiatry*, 27(2), 158-165.

Zadnikar, M., & Kastrin, A. (2011). Effects of hippotherapy and therapeutic horseback riding on postural control or balance in children with cerebral palsy: A meta-analysis. *Developmental Medicine & Child Neurology*, 53 (8), 684-691.

Made in the USA
Monee, IL
05 November 2026

67977834R00052